T0245222

CAMBRIDGE LIBRARY COLLECTION

Books of enduring scholarly value

Perspectives from the Royal Asiatic Society

A long-standing European fascination with Asia, from the Middle East to China and Japan, came more sharply into focus during the early modern period, as voyages of exploration gave rise to commercial enterprises such as the East India companies, and their attendant colonial activities. This series is a collaborative venture between the Cambridge Library Collection and the Royal Asiatic Society of Great Britain and Ireland, founded in 1823. The series reissues works from the Royal Asiatic Society's extensive library of rare books and sponsored publications that shed light on eighteenth- and nineteenth-century European responses to the cultures of the Middle East and Asia. The selection covers Asian languages, literature, religions, philosophy, historiography, law, mathematics and science, as studied and translated by Europeans and presented for Western readers.

A Historical View of the Hindu Astronomy

Shrouded in poetry, the earliest accounts of Hindu astronomy can strike modern readers as obscure. They involve the marriage of the moon to twenty-seven princesses, a war between gods and giants, and shadows that give birth to planets. In this fascinating study, first published in Calcutta in 1823 and reissued here in the 1825 edition, John Bentley (*c.*1750–1824) strives to strip back the mythical aspects of the stories to reveal their foundations. He points out that early Hindu astronomers divided the night sky into twenty-seven sections; that a solar eclipse could have been described as an epic war between light and dark; and that Saturn is often observed in the Earth's shadow. Using data from modern astronomical tables, he dates events, texts and people, whether mythical or factual, as well as charting the history of Indian astronomy from its earliest records to its modern developments.

A Historical View of the Hindu Astronomy

From the Earliest Dawn of that Science in India to the Present Time

John Bentley

CAMBRIDGE UNIVERSITY PRESS

Cambridge, New York, Melbourne, Madrid, Cape Town,
Singapore, São Paolo, Delhi, Mexico City

Published in the United States of America by Cambridge University Press, New York

www.cambridge.org
Information on this title: www.cambridge.org/9781108055420

This edition first published 1825
This digitally printed version 2013

ISBN 978-1-108-05542-0 Paperback

A

HISTORICAL VIEW

OF THE

HINDU ASTRONOMY,

FROM

THE EARLIEST DAWN OF THAT SCIENCE IN INDIA,

TO THE PRESENT TIME.

IN TWO PARTS.

PART I.

THE ANCIENT ASTRONOMY.

PART II.

THE MODERN ASTRONOMY,

WITH AN EXPLANATION OF THE APPARENT CAUSE OF ITS INTRODUCTION, AND THE VARIOUS IMPOSITIONS THAT FOLLOWED.

TO WHICH ARE ADDED,

I.—Hindu Tables of Equations.
II.—Remarks on the Chinese Astronomy.
III.—Translations of certain Hieroglyphics, called the Zodiacs of Dendera.

BY JOHN BENTLEY,

MEMBER OF THE ASIATIC SOCIETY.

LONDON:

SMITH, ELDER, & CO. 65, CORNHILL.

MDCCCXXV.

PREFACE.

THE subject of the present essay was undertaken many years ago; but owing to a variety of causes, which could not be foreseen, its publication has been delayed much beyond the time that was originally intended. The first was, that I found it difficult to obtain any correct information respecting the ancient astronomy of the Hindus, by reason of all the ancient works having been purposely destroyed or concealed, since the introduction of the systems now in use. I was therefore induced, by way of saving time, to lay before the Asiatic Society in A.D. 1799, a paper "On the antiquity of the *Surya Siddhanta,* and the formation of the Astronomical Cycles therein contained," intending to follow it up with the present essay, as soon as the necessary facts and data that I was in search of could be obtained. My paper on the antiquity of the *Surya Siddhanta* was published in the sixth volume of the Asiatic Researches; and I was in great hopes that it would open the eyes of the learned and scientific in Europe, in respect of the real state of the Hindu astronomy, and dispel those prejudices which, from a want of a due knowledge of the subject, have arisen in favour of its supposed extraordinary antiquity. In this hope I have not been disappointed; for some of the first astronomers and scientific men of the age have concurred in the conclusions there drawn respecting the age of the work. The age of that work, however, had nothing to do with the real antiquity of the Hindu astronomy, which was intended to be the subject of the present essay, and would have been long ago published in the Asiatic Researches, had it not been for circumstances, some of which I shall now explain, that totally prevented it. It would appear, that shortly after the publication of the sixth volume of the Asiatic Researches, the harmony of the Society, and the zeal of its members in promoting the object of the institution, were nearly extinguished, by means of certain attacks made on the labours of some of the members, in a periodical work called the Edinburgh Review, apparently with a view of putting down all further researches into antiquity, and the investigation of truth. The consequence was, as might have been easily foreseen — a general apathy and disgust amongst the members, who naturally said to themselves, "If these sneers and scoffs are to be our thanks, it is unnecessary that we should labour in the field any longer: it is better by far that we should refrain from our labour, than involve ourselves in perpetual disputes with persons concealed, who may be both capable and willing to do us an injury, without our being able to ascertain whence it came."

A more fatal blow could not be aimed by the greatest enemy to the institution than had been thus inflicted on it: for, setting aside the general apathy, it would ultimately, by its effects, not only be productive of the greatest evils to the welfare of the Society, but destructive to the intention of its original founder, which was thus perverted; so that, instead of the institution being "for enquiring into the history and antiquities, the arts, sciences, and literature of Asia," as expressed by the title-page of their work, it became of a direct opposite nature. Surely, it ought to have been foreseen, that such a mode of proceeding would ultimately tend to the loss of the Society, and perhaps to its final dissolution. The attack made on any member must obviously be intended to diminish the value of the writings or essays of the member so attacked; consequently it must have the same effect, if it has any at all, in reducing the value of the volume of the Researches. The injury is intended for the individual, but it falls only on the Society; for the individual, becoming disgusted with the treatment he receives at the hands of secret enemies, is obliged to reserve the publication of the result of his labours for some future time, when more liberal sentiments may prevail. In consequence of the attack made on my essay on the antiquity of the *Surya Siddhanta,* I wrote another paper in the eighth volume of the Asiatic Researches, pointing out the complete ignorance of the reviewer, and his ill nature in making an attack, where in fact there was not the most distant foundation for it. I there showed by a table the gradual decrease of the errors in the *Surya Siddhanta,* from the year 3102 before Christ, down to A.D. 999, and also for two periods later, in order to show the increase of the errors again in an opposite direction; thereby showing, that the point at which the errors were the least possible was between seven and eight hundred years back at the time. This table is so plain, that a schoolboy totally unacquainted with astronomy can understand it; but, plain and simple as it is, the reviewer, to show his knowledge of astronomy, I suppose, as well as his ill will, has thought fit to attack it by such sophistry and ignorance as, I believe, never before came from the pen of a reviewer; and am rather surprised that it could be at all admitted into the Review, which certainly did it no good, as casting a reflection on the abilities of those who conducted it. But, in order that others may be able to judge of the truth or falsehood of the assertions of the reviewer, I shall here give the three first columns of the table, which are all that are here wanting to exhibit the decrease in the errors.

THE TABLE.

Planets, &c.	B.C. 3102.	A.D. 499.	A.D. 999.
Moon, ...	5° 52′ 34″ –	0° 20′ 14″ –	0° 1′ 2″ –
——— Apogee,	30 11 25 –	4 52 53 –	1 21 59 –
——— Node,	23 37 31 +	3 56 6 +	1 12 1 +
Venus, ...	32 43 46 –	3 33 41 –	0 29 22 +
Mars, ...	12 5 42 +	2 32 42 +	1 13 8 +
——— Aphel.	9 47 0 +	1 30 50 +	0 21 55 +
Jupiter, ...	17 12 36 –	1 48 56 –	0 24 20 +
Saturn, ...	21 25 43 +	2 50 9 +	0 3 33 –
Sun's apogee,	3 15 53 +	0 5 45 +	0 33 45 –

There is no man, I believe, in his senses, who, on inspection of this table, will not say, that the author of the work must have lived at that period of time when the errors were least, instead of that period when they were greatest; but our reviewer would wish his readers to believe the contrary; and that the author of the *Surya Siddhanta,* instead of living about the year A.D. 1000, when the error in the moon's place was only about 1, lived 3102 before Christ, when the error amounted to near 6°: from which circumstance, it will naturally be concluded by many, that he must either be mad, or entirely ignorant of the nature of astronomy, which science could admit of no such conclusion; because, from the quantity of the error in the moon's place 3102 before Christ, it was impossible that the computations by the *Surya Siddhanta,* could give, or foretel the eclipses of the sun and moon, which clearly shows the absurdity of the reviewer's notions. Lest, however, it might be supposed that I am not speaking the truth, I will here transcribe his own words, which I think will point out still more his sophistry, ignorance, and ill will.

"Let us next consider the criterion which Mr. Bentley himself proposes for determining the age of a system of astronomical tables, from the consideration of the tables themselves, independently of testimony, tradition, or any external evidence. Such a criterion is precisely the thing wanted on the present occasion: but we can by no means approve of that particular one which he endeavours to establish. It is founded on this maxim, that the time of the construction of any set of tables must be that at which they agree best with the heavens. Hence, when such tables are given, and we wish to determine their antiquity, we have only to compute from them the places of the sun and moon, &c. for different times, considerably distant from one another: to compare these places with those given by the best modern tables; and the time when they approach the nearest to one another is to be taken for the time when the tables were constructed."

The reviewer then goes on with his sophistry, to endeavour to set aside this rule, thus:—"As it must be an object, in all astronomical tables, to represent the state of the heavens tolerably near the truth at the time when they are composed, it must be allowed that this rule is not destitute of plausibility. On examination, however, it will be found very fallacious, and such as might lead into great mistakes." The reviewer then proceeds to show his extraordinary skill and sagacity, by saying, that "Astronomical tables are liable to errors of two different kinds, that may sometimes be in the same, sometimes in opposite directions. One of them concerns the radical places at the epoch from which the motions are counted, the other concerns the mean motions themselves, that is to say, the mean rate, or angular velocity of the planet. Of these the first remains fixed, and its effect at all times is the same: the second again is variable, and its effect increases proportionally to the time. If, therefore, they are opposite, the one in excess and the other in defect, they must partly destroy one another; and the one, increasing continually, will at length become equal to the other, when there will of consequence be no error at all; after which the error will fall on the opposite side, and will increase continually. Here the moment of no error, or that when the tables are perfectly correct, is evidently distant from the time of the construction of the tables, and may be very long either before or after that period. Suppose, for example, that in constructing tables of the sun's motion, we are to set off from the beginning of the present century, and that we make the sun's place for the beginning of the year 1801 more advanced by half a degree than it was in reality. Suppose also that the mean motion set down in our tables is erroneous, in a way opposite to the former, and is less than the truth by 1″ in a year. The place of the sun then, as assigned from tables for every year subsequent to 1800, will, from the first of the above causes, be half a degree too far advanced, and, from the second, it will be too little advanced, by as many seconds as there are years: when the number of years become as great as that of the seconds in 30, that is, when it is equal to 1800, the two errors will destroy one another, and the tables will give the place of the sun perfectly exact. Were we, therefore, to ascertain the age of the tables by Mr. Bentley's rule, we should commit an error of 1800 years; from which we may judge of the credit due to that rule, as a guide in chronological researches."

Here the reviewer's plausible sophistry may be clearly seen through. He assumes an error in the sun's place at the epoch of the tables of 30″, and an error in the mean motion of 1″ per year. All this, we see, is upon one side, and a mere assumption. Have I not the same right to assume a similar error in one of the planets, that is, an error of 30′, in its radical position, and an error of 1″ per annum in its motion, so that in 1800 years the error of 30′ would be cancelled? Now if this error so assúmed be of an opposite nature to the one he assumed, the one cancels the other, and the epoch of the tables would not be at all affected by the circumstance: his conclusion, therefore, in regard to my rule is incorrect; for I do not determine the age of any system of astronomy by a single item, as he thought proper to assume on this occasion, to give the greater plausibility to his assertions; I make use of as many as appear to be the most correct, because the errors counteract each other. Why did he not apply the rule to some tables, the age of which was known to him, the same as I have done in respect of the system of Āryabhatta, in the third section of the second part of my essay, the author of which gives his own date? I have also applied the same rule to the *Brahma Siddhanta*, the first of the modern astronomical works introduced in A.D. 538, to which the reader may refer, and where he will find the method by which the system was constructed explained at length. The application of the rule, in both these instances, when we had the actual dates of the system before us, demonstrate that it is perfectly just, so far at least as we require it. I shall, however, for the sake of exposing the fallacy of such arguments, apply the rule to the determination of the age of certain well known modern tables, which I hope will put this cavilling at rest, or expose the author of them to the censure he deserves.

A set of astronomical tables being put into my hands, to determine their antiquity, I accordingly compared them with the third edition of La Lande's, which is supposed to be correct, and find the errors gradually diminishing from the Christian era down to about the year A. D. 1744, as in the following

TABLE.

Planets, &c.		A. D. 0	A. D. 1000	A. D. 1500	A. D. 1700	Year when no Error.	Years Selected.
						A	B
Sun,	+	0° 1′ 26″	0° 0′ 42″5	0° 0′ 19″7	0° 0′ 11″	1948	
Sun's apogee	–	1 37 2	0 41 8	0 13 11	0 1 59	1737	1737
Moon,	–	0 6 38	0 2 48	0 0 53	0 0 7	1730	1730
— Apogee,	+	0 0 58	0 0 58	0 0 58	0 0 58		
— Node,	+	0 0 51	0 0 51	0 0 51	0 0 51		
Mercury,	–	2 18 29	1 0 9	0 20 59	0 5 20	1768	1768
— Aphel.	–	6 59 27	3 0 17	1 0 42	0 12 54	1754	1754
— Node,		0 49 2	0 20 42	0 6 32	0 0 53	1731	1731
Venus,	+	0 12 31	0 4 31	0 0 31	–0 1 5	1565	
— Aphel.	–	49 0 8	20 50 8	6 45 8	1 7 8	1740	1740
— Node,		0 0 0	0 0 0	0 0 0	0 0 0		
Mars,	+	0 0 30	0 0 30	0 0 30	0 0 30		
— Aphel.		0 0 0	0 0 0	0 0 0	0 0 0		
— Node,		5 46 15	2 29 35	0 51 15	0 11 55	1761	1761
Jupiter,	–	2 23 17	0 53 49	0 4 5	+0 15 49	1541	
— Aphel.	–	2 32 14	1 4 24	0 20 29	0 2 55	1733	1733
— Node,	–	11 27 6	4 42 6	1 19 30	+0 1 23	1696	
Saturn,	+	4 17 53	1 26 49	0 1 17	–0 32 56	1507	
— Aphel.	–	9 52 35	3 13 45	+ 5 40	1 25 26	1486	
— Node,	+	0 47 7	0 21 17	0 8 22	0 2 12	1823	
						A	B
Sum of the years in the two last columns, ...						25520	13954
The same divided by the number of items, the mean results are						1701	1744¼

To make my meaning better understood, I have selected into the last column all those years that agree near to each other, and the result from thence comes out 1744, the age of the astronomical tables: but to show how far the errors in the years in the other column would affect this result, I have cast them up as they stand, and they give then 1701 years. The notion of the reviewer is here shown in the item of 1948, at the beginning of the column under A, in order to show his error; for that item is balanced by others on the contrary side:—but is not the idea of using an erroneous result, when we have so many others to choose from, truly ridiculous? Surely the astronomer must be at liberty to employ those results he finds most correct, and not those that appear erroneous. The reviewer's ideas in this respect are totally unscientific, and rest upon nothing but sophistry, misrepresentation, and deception.

No result can be drawn when the motions are the same by both tables, as in the moon's apogee and node, the node of Venus, Mars and his aphelion; for the errors neither increase nor diminish. The tables here given as an example to the rule, are those of La Lande, first or second edition; but the title-page and date are wanting. The reason why the mean result makes the tables older than the epoch, which I believe was 1770, is, that the observations on which they are founded are anterior to it. None of the Hindu artificial systems are ancient: they are all since A. D. 538, the year in which the modern astronomy commenced, and not at the beginning of the *Kali Yuga,* as imagined by Bailly or others; for they had no astronomy then, as I have fully proved in the Essay, where I have brought forward all the astronomical facts and observations that could be collected relative to their ancient astronomy, showing that it did not go further back than about 1425 before Christ, which was only the dawn of astronomy in India.

The reviewer must be greatly mistaken, if he imagines that the motions of the planets in Hindu artificial systems, were drawn from two actual observations, as in the European methods, and " that their merit, and their claim to antiquity, was decided"—" by the accuracy of the mean motions, as contained in the tables." Now the assertion thus boldly made is not correct; for the Hindu artificial systems, and tables drawn from them, which give the motions alluded to, are not constructed on the European principle. The European method requires that the motions be determined by two actual observations at least, and at a considerable distance of time: the Hindu, in artificial systems, requires no such thing; he makes use of but one observation, which is in the time of the observer: the other he assumes; that is to say, he assumes that the planets were in a line of mean conjunction at the beginning of the *Kali Yuga*, and from thence draws the motion that would give the position of the planet in his own time, agreeing with his observation. To make this matter plain, I will give an example.

Suppose that in A.D. 939, at the end of the year 4040 of the *Kali Yuga*, there were two astronomers making observations on the planet Venus at the same place, one being a European, the other a Hindu, and that they both found the place of the planet in the Hindu sphere to be then $2^s\ 19°\ 55'\ 12''$. The Hindu astronomer says to the European: "We must now find the mean annual motion that will give this position; and observes, that at the beginning of the *Kali Yuga*, the planets were all in a line of mean conjunction in the beginning of the Hindu sphere; consequently, that the mean annual motion of Venus, multiplied by 4040, the years then elapsed, must give the position, and therefore the mean motion must be $= 7^s\ 15°\ 11'\ 52''8$; for if this quantity be multiplied by 4040, it will produce $2^s\ 19°\ 55'\ 12''$." The European astronomer observes: "We have agreed in the actual position of Venus at the end of the year 4040 of the *Kali Yuga*, because it depends on actual observation made now by both of us; but the assumption that the planets were in a line of mean conjunction in the beginning of the *Kali Yuga*, I cannot admit to be true: for by our tables, which I take to be perfectly correct, Venus was not in the position assumed by you, for her longitude then, was $1^s\ 2^\circ\ 43'\ 46''6$; consequently the mean annual motion must be less than what you make it, and by my calculation comes out exactly $7^s\ 15^\circ\ 11'\ 23''635$, which, if multiplied by 4040 years,

we shall have . . .	$1^s\ 17^\circ\ 11'\ 25''4$
Add position of Venus at the epoch,	1 2 43 46 6
We get her mean longitude now,	=2 19 55 12 "

The Hindu astronomer replies: "Yes, Master European, your mode may be true; for we have no means of detecting its errors, since we had no observations at the beginning of the *Kali Yuga:* we can only say, that you take more trouble by adding the supposed position of Venus at the epoch than we do. We manage it otherwise, as you see, by simply taking the mean motion such, that it gives us the position without any addition or subtraction, which labour we save."

The European astronomer says: "You certainly save both labour and time; but still I do not approve of your method, because it is incorrect: for though it gives the same longitude to the planet as mine does for this moment of time, yet it will not continue to do so for any length of time. For instance, next year there will be a difference between us of $29''165$, and it will for ever after continue to increase at that rate yearly. But this is not all: our great astronomers, who may live between eight and nine hundred years hence, will be deceived by the annual motions which you thus deduce, some being greater, and others less than we give them, thereby, according to physical ideas, indicating great antiquity; so that your time will be thrown back between two and three thousand years." The Hindu astronomer replies: "I am very glad to hear it; I did not mean deception; but since you will have it so, take it. I did not think the European astronomers to be such fools as to be deceived by our manner of deriving the mean motions."

Thus I have endeavoured to show, in a familiar way, the Hindu method of deriving the mean annual motions, which being totally different from the European, renders the method of Bailly and others, of ascertaining the supposed antiquity of astronomical tables from the quantity of the mean motions, as totally inapplicable. But it may be done, if we apply the principle upon which they have been derived; otherwise not. Thus, suppose we have the mean annual motion of Venus as above given $= 7^s\,15^\circ\,11'\,52''8$, to find what year this motion corresponds to, that is, the year in which it would give the mean longitude of Venus corresponding to observation, the mean motion of Venus, according to Europeans, in the Hindu sphere, would be as above $= 7^s\,15^\circ\,11'\,23''635$

Which taken from the former, 7 15 11 52 8

Leaves a remainder of . . 29 165

The position at the epoch of the *Kali Yuga*, according to Europeans $= 1^s\ 2^\circ\ 43'\ 46''6$; divide the latter by the former, and we have $\frac{1^s\ 2^\circ\ 43'\ 46''6}{\cdot 29''165} = 4040$, the year from the epoch when it gives the position of Venus agreeing with observation.

The reviewer says: "The antiquity of it (the *Surya Siddhānta*), has been conceived to be very great, as it is reckoned the most ancient astronomical treatise of the Hindus; but, according to Mr. Bentley, that antiquity extends to no more remote period than the year 1068 of our era. The main argument on which this determination is founded, seems to us subject to considerable difficulty. It supposes what is by no means certain, that the Hindu astronomers deduced the mean motions of the planets from a comparison of a real observation with one that was purely fictitious. This is no where proved by Mr. Bentley, though taken as the basis of all his computations." The assertion of the reviewer in this instance is positively untrue; for I have shown, from the data given in the *Surya Siddhānta* itself, that all the mean motions and positions of the planets given by that work are expressly deduced from the assumption of the planets being all in a line of mean conjunction in the beginning of the Hindu sphere, at midnight, at the beginning of the *Kali Yuga*, and on the meridian of Lanka. These data are all derived from the work itself: and what will show it to be assumed, and not a real epoch of observation, is, that the vernal equinoctial point, or the beginning of Aries in the tropical sphere, was then assumed to be also in the same point with the planets, which we know could not have been the case; for the sun's mean longitude in the European sphere at that moment was $10^s\ 1°\ 1'\ 1''$, and therefore 60 days short of the time: yet, notwithstanding all this apparent absurdity, it is the epoch from which not only the motions of the planets are reckoned and drawn, but also the precession of the equinoxes. It is really ridiculous to see a man like the reviewer, who seems to know nothing whatever of the Hindu astronomy, talk on a subject he is unacquainted with, and pretend, with the utmost gravity, that the Hindus, like all other people, must have two or more observations made at a distance, from whence they drew the motions of the planets. They might have hundreds or thousands of observations in ancient times at a distance from each other, and draw the mean motions from thence for other books, but not for the artificial systems now in use, nor the *Surya Siddhānta*, which is entirely on an artificial plan: and all those that are on an artificial plan have been introduced since A.D. 538, for the purpose of imposition, in order that their history and astronomy should be considered by the ignorant as excessively ancient; in which imposition they have certainly but too well succeeded. The example I have above given respecting Venus, shows how the motions are deduced, the position in the astronomer's time being known from observation: and it also shows, from the motion alone, how to determine the time to which it refers. They are the same as are given by the *Surya Siddhānta.*

The Hindus have many astronomical books not on the artificial plan: they are, however, all modern, and do not fall within the scope of my observations, as they are hardly worth noticing. It is to the exposure of the impositions introduced by the artificial systems, that my whole attention has been principally directed.

Is it not most strange that Mr. Bailly, or any other person pretending to a knowledge of astronomy, should place the age of a work at that period of time when its rules made an error in the moon's place of near 6°, in the moon's apogee upwards of 30°, and in the moon's node near 24°? How was it possible for eclipses to be calculated by the rules of that book, when the moon, at the actual time of an eclipse, would be 6° distant from the sun by the rules? And how was it possible for it to point out the precession of the equinoxes, when it erred 60 days in the time? The tables of *Trivalor* and *Chrishnaborum* are from the *Surya Siddhānta,* and of course contained these errors in them for the beginning of the *Kali Yuga;* but Mr. Bailly having imagined that the mean motions were drawn according to the European method, which has been repeatedly shown was not the case, he fell into the mistake.

How is it possible that a man, pretending to a knowledge of the principles of astronomy, should or could give in to such errors? The eclipses of the sun and moon are the most material evidence for determining the date of astronomical tables, whether the motions be drawn on an artificial plan or not: it is by them that the astronomer proves the truth of his tables, and his own abilities; and it is by the time, quantity, and other circumstances of an eclipse, that he is enabled to see and correct, from time to time, the errors that may be concealed in his tables or rules, before he makes them known. Why did not the reviewer determine the age of the *Surya Siddhānta* by this criterion, if he did not like the rule we proposed, in respect of the position and motion of the planets? It may be that he thought the labour of calculating eclipses, and comparing them with those deduced from correct European tables, too great, and, moreover, that it would prove too much, viz. that the *Surya Siddhānta* was not composed 3100 years before the Christian era.

The reviewer, not satisfied with what he says in the tenth volume of the Edinburgh Review, on the subject of my reply to his strictures on my first paper, accuses me of attacking the opinions of Bailly and others, thus:—"Mr. Bentley having with great courage brought forward his own peculiar views, in opposition to the authority of such celebrated names as those of Bailly, Le Gentel, Playfair, and Sir William Jones, it certainly did not occur to us that we could be guilty of any very unpardonable presumption, in venturing to doubt whether his speculations were in all respects conclusive. Mr. Bentley, however, has thought fit to resent our scepticism with a good deal of philosophical warmth, and with unmerciful severity accuses us of both attachment to system, and of relinquishing that system."

The reviewer, not only here, but in every other instance, endeavours to cloke his attack, and support his arguments under the authority of such names as Bailly and others, because I differed from them in opinion. I could not help feeling warm at a wanton and insidious attack being made on me for explaining the nature of the astronomical system contained in the *Surya Siddhānta*, and the formation of the numbers and revolutions it contained, which clearly pointed out that Bailly was completely mistaken in the ideas he had adopted. This was the crime for which I was attacked; and I was the more vexed at it, because it appeared to be done with a view to put down all such investigations for the future, and I was actually so told. So then, if a great man is to commit a mistake, we are not at liberty to point it out; it must remain so for ever, at least the reviewer would so insinuate; but I am of a very different opinion: the greater the man, the more necessary it is to point out his errors, and the foundation of them, that others may not fall into the same. It was for this reason that I pointed out the cause why the motions given in the *Surya Siddhānta*, though a modern work, must of necessity differ from the European, in consequence of the position assumed at the beginning of the *Kali Yuga* being $0^s\ 0^\circ\ 0'\ 0''$. The example I have given above in respect of Venus, explains this circumstance sufficiently clear; so that it is not necessary to insist more on it here. Mr. Playfair, who supported Mr. Bailly with his calculations, was then living: did he consider that I made an attack on him, because I explained the cause of the errors, by which all his calculations became of no use? Most certainly not: though he was wrongfully stated by some as the author of the review, in order to throw the odium upon him, and take it from the real person. I sent Mr. Playfair a copy of my paper on the antiquity of the *Surya Siddhānta*, to open his eyes as to the foundation of Mr. Bailly's mistake: and after the review on it came out, it being industriously fathered on Mr. Playfair, I directed enquiry to be made at Edinburgh through some of Mr. Playfair's most intimate friends, to ascertain from himself if he was the author of the review. The reply was, what I would have expected from a man of candour and science, that he was not the author of the review, and that he could not, consistently with his character, be the author of any such nonsense. What further information he afforded, need not now be noticed.

Having thus far ascertained that Mr. Playfair was not the author of the review in question, nor of those that followed on the same topic, I was anxious to know the opinions of others on the subject: for though I was perfectly satisfied I was right in the conclusions I drew, yet to have the ideas of others, whose skill and knowledge in astronomy could not be doubted, would be highly gratifying. I therefore collected together various astronomical facts, particularly those above alluded to respecting the moon's place, &c. at the beginning of the *Kali Yuga*, in order to show the error, not only of Mr. Bailly, but of the reviewer, who imagined, that by using his name he could do wonders. These facts I forwarded to a friend in London, desiring him to show them to the Astronomer Royal, the late Rev. Dr. Maskelyne, and to get his opinion thereon; which he accordingly did, and transmitted me Dr. Maskelyne's answer, in a letter under date the 12th April, 1811, which I shall now take the liberty to introduce. He says: "I showed your astronomical letter to Dr. Maskelyne; indeed I left it with him several weeks: he returned it to me at his own table, at dinner, with the following observations: —

"I think Bentley right: he has proved by his calculations that there were no real observations made at the beginning of the *Kali Yuga*. Bailly was a pleasing historical writer; but he had more imagination than judgment, and I know that he was condemned by his friends La Lande and La Place, as a *superficial astronomer*, and a very *indifferent calculator*. These two gentlemen entertained the same opinion with myself, with respect to the antiquity of Hindu astronomy; and I think Mr. Bentley has made out satisfactorily the real antiquity of the *Surya Siddhānta*."

It is well known that Dr. Maskelyne was an astronomer of the first-rate abilities, and of the utmost integrity. Here he gives his opinion free from any prejudice. He does not condescend to notice the reviewer, whom, for his sophistry, he considers as beneath his observations.

Delambre, one of the greatest modern astronomers, has also taken up the subject in his Ancient Astronomy; and though he notices the reviewer, he treats his notions with contempt, by deciding at once against the pretended antiquity of the *Surya Siddhānta*. So far, therefore, I thought it right, and in justice to myself, to exhibit the opinions of far superior authorities to an obscure pretender, who dares not to come forth with his own name, knowing, that what he asserts is not true, and that the whole of his object is deception, sophistry, and the misrepresentation of facts. He has complained, that I have attacked him with severity: but whose fault is that? Why has he meddled with me? and that, too, under the mask of concealment, under the name of reviewer, by which he thought to stab me in the dark, to destroy my labours, and to do me every possible injury, without my knowing to what hand I had been indebted for such unprovoked and unexpected usage? Can it be supposed that I would tamely submit to be thus treated, without speaking my mind, and exposing the malice and ill will of the individual who could shamefully, and in spite of truth, act in this manner? What serves to mark his malice and ill will, is his attempt to magnify imaginary defects in my method, while at the same time he endeavours to uphold the method of Bailly, which of all others was the most imperfect. For the quantities of the motions of the planets were totally inapplicable, from their not being derived in the European manner, on which he reckoned. And though the greatest equations of the planets were not liable to the same objection, yet they were too ill determined in Hindu books to answer the purpose. Not satisfied with this marked ill will, he crowns it by accusing me of having attacked Sir W. Jones, Mr. Playfair, and others, thinking thereby, no doubt, as he found his sophistry and misrepresentations not sufficient to answer his views, that their friends would join him in raising a hue and cry against me.

In this, however, he has been disappointed. I certainly attacked no one; but I explained the nature and principles of the modern Hindu system of astronomy, showing that it was entirely contrary to the view that had been taken of it: so that I think I have fairly and clearly proved, that the sophistry and misrepresentations of the reviewer are founded in malice.

By his attempt to uphold the antiquity of Hindu books against absolute facts, he thereby supports all those horrid abuses and impositions found in them, under the pretended sanction of antiquity, viz. the burning of widows, the destroying of infants, and even the immolation of men. Nay, his aim goes still deeper; for by the same means he endeavours to overturn the Mosaic account, and sap the very foundation of our religion: for if we are to believe in the antiquity of Hindu books, as he would wish us, then the Mosaic account is all a fable, or a fiction.

When our just endeavours to do all the good in our power, to stem the torrent of imposition, and to lay the same open to full view, are opposed by secret means, or by persons in concealment counteracting our intentions, we cannot help feeling a regret that such things should exist: that they do exist, however, is certain, and has been fully proved by the preceding pages; and were I so disposed, I could exhibit a great deal more. However, for the present, I must draw a veil over them, and proceed to what may be deemed of more importance; that is, some account of the present Essay.

This work, as I noticed in the beginning of the preface, I began many years ago, in hopes that by giving a clear and concise historical view of the Hindu astronomy, from the earliest period of time in which the science was known, it would contribute greatly to the dispelling of that mist of ignorance under which it had lain so long. In conformity, therefore, with this plan, I begin the first section with the earliest observations known or noticed in Hindu books, which will clearly show that the Hindus had no astronomy, at least that we or themselves know of, anterior to the year 1425 B.C. when it is supposed the Lunar Mansions were formed, and the first observations made.

About the year 1181 B.C. they with great ingenuity formed the months, and gave them names, derived from the Lunar Mansions in the manner explained in the first section, which puts a stop to the imaginary antiquity of all Hindu books and systems that mention the names of the months, let their pretensions be whatever they may.

In the second section, I give the epoch of Rama, deduced from the positions of the planets at his birth, which is confirmed by the eclipse of the sun, and other circumstances, at the churning of the ocean, or war between the gods and the giants; as also the eclipse of the sun, and positions of the planets, at the time his father wished him to join in the government; so that there is not a point in history determined with more certainty and precision than the celebrated epoch of Rama, which may be of some importance to those who make the Hindu history their study, as it will enable them to correct and settle other points by the number of reigns, either before or after Rama, with more certainty than they otherwise could do. I have also noticed the observations then made in respect of the length of the year, the precession of the equinoxes derived from the lunisolar period then discovered, which was the foundation of the changes made in the commencements of the year from time to time. The war between the gods and the giants I have given at full length, in order to show the time to which it referred, and followed it up in the third section by a description of the war between the gods and the giants in the west, with all the circumstances I could find relating to it, for the better determining the time; which seems to be of importance for establishing with more certainty the epoch of the formation of the constellations, the Argonautic expedition, as it is called, and the time of Hesiod, who gives a description of the war, which could not, therefore, have been written till the close of it, which I have placed in the year 746 B.C.

In the fourth section, I have given the epoch of Yudhishthira, Parasara, and Garga, which is a very material point in Hindu history. I have explained the nature of the term, the *Rishis* in *Maghā,* as introduced by the astronomers of that period; which term the moderns have entirely perverted, to answer their own impositions. I have also explained some other passages of Parāsara and Garga, relative to the positions of the colures, which likewise have been perverted in modern times.

In the fifth section, I have given the four periods into which the Hindu history was divided in ancient times, that is, as early as the year 204 before Christ, when this division seems to have been first invented. It is very remarkable, that by this arrangement, the creation took place at the very year of the Mosaical flood, by which it appears they had then no knowledge whatever of any history anterior to that circumstance. It serves, however, I think, as a proof of the year of the deluge being correctly given, as the Hindus must have preserved it by tradition as the year of the creation, which was very natural.

In the sixth section, I have given the nine patriarchal periods, called *Manwantaras.* These appear to have been invented in the first or second century of the Christian era, with a view, I believe, of correcting the error in respect of the creation, as given by the four ages. For the first of these patriarchal periods goes back to the year 4225 B.C. which is called the creation. By the former division, the creation took place at, or near the vernal equinox: by the latter, on the 25th of October, at the autumnal equinox. But though the latter altered the time of the creation, it was only for man; for the animals they stated to have been created at a much later period.

In the Modern Astronomy, first section, I have given a full view of the introduction of the modern systems, by which the creation has been thrown back into antiquity several millions of years. I have shown, by operations at length, how the epoch of the modern *Kali Yuga* was settled, and the method by which the planetary motions, positions, &c. were adapted to the system of Brahma, to answer the purpose in view. These systems have been the origin of a great part of the modern impositions, which would be too long to describe here.

In the second section, I have given the system called *Varāha*, as given in the *Surya Siddhānta*, &c. and shown the object of it was to support the former system in imposition. Its date I have shown by computation, and explained the ingenious contri vance of the author for calculating with ease the precession of the equinoxes, &c.

In the third section, I have given the system of Āryabhatta, together with that of Parāsara, which was framed by him for the purpose of imposition, the nature of which I have fully explained. The date is given by Āryabhatta himself, which is also corroborated by computation made from his system.

I have also noticed his geometry, &c. which appears to have been the same, nearly, with what is given in the *Lilāvati* by Bhāskara, who wrote a commentary on Āryabhatta's work, which probably was the foundation or origin of the *Lilāvati*, with a few more modern improvements added.

In the fourth section, I have noticed Varāha Mihira, and computed his time from the heliacal rising of Canopus, as given by himself, in his *Sanhita*, which was when the sun was 7° short of Virgo, or 23° of Leo; which computation makes his time the same as given by himself in the *Jātakārnava*, in which he gives the positions of the aphelia of the planets for the year *Saca* 1450, or A.D. 1528, which, therefore, makes him contemporary nearly with the emperor Akber. He was one of those who endeavoured to assist in the modern impositions; for he attempted to pervert the meaning of the epoch of Yudhisht hira, which was the year 2526 of the modern *Kali Yuga*, by saying, that the meaning was, that he lived that number of years before *Saca*. He also supported, for the same purpose, the idle notion first introduced by Āryabhatta, about the motions of the *Rishis* being one lunar Mansion in one hundred years — a thing in itself too absurd to be noticed by an astronomer, unless for the purpose of imposition.

Thus, from the above date, 2526, he makes the *Rishis* to be in *Maghā* in the year of *Kali Yuga* 653
Lalla, who follows him, gives . . . 614

Muniswara gives 600
Āryabhatta, the first impostor, . . 663
System of Parāsara by Āryabhatta, . 666

And upon one or other of these fictitious eras of Yudhishthira, the modern histories of Cashmere, and other parts of India, are now erected, and given to the world as true: the whole of which, however, is shown to be false, from the positions of the planets in the time of Garga, 548 B.C.; the epoch of Rama, 961 B.C.; the epoch of the formation of the months, &c. &c. which overthrow the whole imposition.

In the fifth section, I have shown the cause of Varāha Mihira and Bhāskara Āchārya being thrown back into antiquity, to have arisen from a trick played on Akber. I have shown the various means that were adopted to support the imposition, by interpolations and forgeries of every description; which system of forgeries and impositions has continued down to the present day unabated, nay, rather with many new additions and improvements.

In the sixth section, I have been obliged to come forward in my own defence, against an extraordinary mode that has been adopted by Mr. Colebrooke, for opposing my computation on the antiquity of the *Surya Siddhānta*, and for throwing back into antiquity Varāha Mihira, and all others who state the solstices to be in the beginning of Cancer and Capricorn. The first he endeavours to effect by means of the tables of Lunar Mansions, imagining thereby that the book must be as old as the time to which the positions of the stars refer. But this is not the case: the tables are found in books of all ages, and are inserted in them merely as tables of reference, having no connection whatever with the age of the book in which they are inserted, as fully proved by the books themselves. The second he endeavours to effect, by giving the names Aries, Taurus, Gemini, &c. to the signs or divisions of the Hindu sphere, beginning from the lunar asterism *Aswinī*, which is completely proved by all Hindu books extant, as well as by a translation of his own from Bhattotpala, to be not only erroneous, but inconsistent with the Hindu astronomy, which assigns these names to the signs of the ecliptic, beginning from the vernal equinoctial point, and to no other signs whatever.

As nothing more seems to me to be necessary towards understanding what I have written on the Hindu astronomy, I shall now close this preface, with recommending and consigning my labours in the investigation of truth, to the friendly protection, care, and attention of all liberal and unprejudiced men of science, as this will in all probability be the last effort I shall make: and I fear that there are but few inclined to follow it up, as they would receive no thanks for their pains, but, on the contrary, opposition and ill will, the only rewards which I have met with for my labours.

TABLE OF CONTENTS.

PART I.

SECTION I.

Page

The early part of the Hindu astronomy involved in great obscurity—The Lunar Mansions the most ancient part of the Hindu astronomy—The time they were formed determined from astronomical data—Second epoch in the Hindu astronomy—Many improvements made—The solar months formed, and named on astronomical principles—Tropical Lunar Mansions introduced—The seasons of the year marked, and fixed to the tropical revolutions of the sun—All explained by a plate, and the time demonstrated—Birth of the goddess Durgā, or the year, with the month *Āswina*, 1

SECTION II.

Rāma—Time of his birth determined from astronomical data—Date of the *Rāmāyana* determined—Churning of the Ocean, or War between the Gods and the Giants, description of it—Time of it determined from astronomical and other data—Birth of Saturn—The month *Kārtika* made the commencement of the year—The rate of precession determined—A lunisolar period discovered, on which were founded the changes to take place in the commencement of the year—A table of eight periods of the same, and the year of their respective commencement, &c. 14

SECTION III.

The War between the Gods and the Giants in the West, described by Hesiod in his Theogony—The time of it determined from various data—Its duration ten years and five months—The zodiacal constellations and others then formed—The original idea of some of them appears to have been derived from the Hindus—The tropical signs named after the zodiacal constellations—The months named after the tropical signs with which they then coincided—Their names compared—Homer and Hesiod not so ancient as generally supposed—The year represented by the ancients under a great variety of names and personifications, as Mercury, Hermes, Anubis, Buddha, &c. 36

SECTION IV.

Commencement of the third astronomical period—The precession then—The term *Rishis* in *Maghā* explained—Parāsara and Garga cited—The heliacal rising of Canopus in the time of Parāsara—The same computed—Positions of the planets when Garga wrote—The time deduced—The real epoch of Yudishthira, 2526 of the Kali Yuga, perverted by the moderns—The fourth astronomical period, ... 64

SECTION V.

Page

Commencement of the fifth astronomical, period — Astronomy further improved — More accurate tables formed, and equations introduced — The Hindu history divided into four periods, and the commencement of each settled astronomically — Tables of the four ages, and their respective years of commencement, with the errors in the tables then used — Commencement of the sixth astronomical period — Christianity preached in India by St. Thomas — The Hindu history divided into new periods, called *Manwantaras*, and, like the former, settled on astronomical principles — Table of the periods, with the errors in the tables used — These occasion no derangement in the Hindu history; only they carry back the creation to the year 4225 B.C. 25th October, at the autumnal equinox — The seventh astronomical period commences, the end of which terminates the ancient astronomy of the Hindus, A. D. 538 74

PART II.

SECTION I.

Commencement of the eighth astronomical period, the beginning of the modern astronomy — The Brahmins introduce new and enormous periods into their history — The means adopted on the occasion — The new periods explained — Fixed by astronomical computations, the nature of which is explained at length — The revolutions of the planets determined, and adjusted to the system of years so introduced — Method of determining the antiquity of the system, supposing the same unknown — The same by a table of errors continually decreasing down to the epoch — The positions of the stars given in the Hindu tables explained with a diagram — Table of the Lunar Asterisms — The names of the signs Aries, Taurus, &c. introduced from the West, and still used to represent the signs as beginning from the vernal equinoctial point — Some of the impositions of modern commentators and others noticed — The system intended as a blow against the Christians — The *Avatars* invented for the same purpose — Krishna the *Avatar* noticed — His nativity computed from the positions of the planets at his birth, 83

SECTION II.

System of Varāha — Framed in the ninth century — The object of it — Works in which it is given — Observation on Canopus referring to A. D. 928 — Revolutions of the planets, &c. in the system — Years elapsed to the beginning of the *Kali Yuga* — Formation of the system, with remarks — Compared with the system of Brahma — Age of the system determined — Lunar Asterisms — The places of some stars not agreeing with the names of the mansions — The cause explained, and shown in a table — Precession of the equinoxes — The method employed artificial, by assuming the motion in an epicycle — Explained by a diagram — The terms libration or

Page

oscillation inconsistent with the author's meaning, which is further explained by the commentator, &c. &c. 115

SECTION III.

The *Ārya Siddhanta*, by Āryabhatta — Its date A. D. 1322 — The object of it — The system it contains — Its formation — Precession of the equinoxes — Mode of computing it — The *Rishis* — The object of introducing them, and the manner of computing their place — The *Parāsara Siddhanta*, and the object of the author in exhibiting it — The system of the *Parāsara Siddhanta* — The computation of the *Rishis* by this system — Age of the *Ārya Siddhanta*, confirmed by computation from astronomical data — Age of the system of Parāsara determined — Found to be of the same age with that of Āryabhatta — Latitudes and longitudes of the stars — Geometry of Āryabhatta same as Bhāskara's — His rules for showing the proportion of the diameter of a circle to its circumference, &c. &c. 138

SECTION IV.

Varāha Mihira, like Aryabhatta, endeavours to support the new order of things — Perverts the meaning of a passage relating to the epoch of Yudisht'hira, whom he places 2448 before Christ — Varāha Mihira mentions the *Surya Siddhanta* and Āryabhatta — States the heliacal rising of Canopus at *Ujein*, when the sun was 7° short of Virgo — Gives the positions of the aphelia of the planets in the *Jātakarnava* for the year 1450 Saca, or A. D. 1528 — The heliacal rising of Canopus at *Ujein* computed for that year, the result agrees with that which Varāha stated, being 7° short of Virgo — The point of heliacal rising of Canopus at *Ujein*, in the Hindu sphere, shown, ... 157

SECTION V.

The cause of Varāha Mihira being thrown back into antiquity by the moderns explained — The reason of two Varāha Mihiras and two Bhāskaras explained by the imposition on Akber — Bhāskara thrown back to A. D. 1150 — A number of forgeries to support the imposition — Spurious *Arya Siddhanta* — Two *Bhasvatis* — Pretended ancient commentaries — Interpolations — The *Pancha Siddhantika* — False positions of the colures — Artificial rules for the cosmical rising of Canopus by the *Bhāsvati* — By the *Pancha Siddhantiki* — By Kesava — By the *Graha Lāghava* — The time to which they refer appears to be about the middle of last century — The heliacal rising of Canopus by the *Brahma Vaivarta*, and *Bhavisya Puranas*, when the sun was 3° short of Virgo — A view of the impositions arising from spurious books — Laksmidasa, a commentator on the *Siddhanta Siromani*, pretended to be a grandson of Kesava, and to have written in A. D. 1500 — Determines the cosmical rising of Canopus at Benares for that year — The spurious *Ārya Siddhanta* examined, and shewn to be a modern forgery — The system it contains, how framed — Gives the proportion of the diameter of a circle to its circumference the same as Bhāskara — Quotes the *Brahma Siddhanta*, Brahma Gupta, and the

Page

Surya Siddhanta — The *Pulisa Siddhanta*, another forgery, noticed — Forgeries of books innumerable — The *Brahma Siddhanta Sphuta*, another forgery — The object of the forgery — The spurious *Brahma Siddhanta* quotes the spurious *Ārya Siddhanta, Pulisa Siddhanta*, and Varāha Mihira, thereby proving it to be a forgery, and, perhaps, by the same individual — Quotation made from it to show the same — Mistake about the positions of the colures, and the meaning of the terms Aries, Taurus, &c. — Interpolations — Vishnu Chandra, &c. 164

SECTION VI.

Self-defence the object of this section — The notions of Mr. Colebrooke respecting the positions of the stars at the general epoch, as indicating the age of the works in which they are found, inconsistent with real facts, being given in books of all ages — Mr. Colebrooke's notions respecting the names Aries, Taurus, &c. being applied to the signs of the Hindu sphere, incorrect — Proved to belong exclusively to the signs of the tropical sphere, by tables and passages in modern Hindu books — Passage from Brahma Gupta to the same effect — Passage from Varāha Mihira to the same effect — Another from the *Tatwachintamani*, containing a computation of the sun's place reckoned both from *Aswinī* and Aries, to the same effect — A translation of a passage in Bhattotpala's commentary to the same effect — The translation by Mr. Colebrooke himself, but not published or noticed by him — Nor the other facts stated — Mr. Colebrooke notices the heliacal rising of Canopus at *Ujein*, when the sun was 7° short of Virgo, mentioned by Varāha Mihira, but does not tell us the time to which it refers — Notices other risings, but without reference to time — Mistaken with respect to the time of rising of Canopus in the time of Parāsara — A passage from Garga explained — Three periods of Canopus's heliacal rising — The 8th and 15th of *Aswina*, and 8th of *Kārtika*, mistranlsated by Mr. Colebrooke — The true meaning given — The time to which they refer explained in a note, ... 196

APPENDIX,

I.—Hindu Tables of Equations, &c. for calculating the true Heliocentric and Geocentric Places of the Planets, &c. 217

II.—Remarks on the Chinese Astronomy, proving, from their Lunar Mansions, that the Science is much more modern among them than is generally believed. The names of their Constellations are added, with the Stars in each. 231

III.—Translations of certain Heiroglyphics, which hitherto have been called (though erroneously) the Zodiacs of Dendera in Egypt... 249

PART I.

THE ANCIENT ASTRONOMY

OF

THE HINDUS.

SECTION I.

FROM 1425 TO 961 B.C.

The early part of the Hindu Astronomy involved in great obscurity—The Lunar Mansions the most ancient part of the Hindu Astronomy—The time they were formed determined from Astronomical Data—Second epoch in the Hindu Astronomy—Many improvements made—The Solar Months formed, and named, on Astronomical principles—Tropical Lunar Mansions introduced—The Seasons of the Year marked, and fixed to the Tropical Revolutions of the Sun—All explained by a Plate, and the time demonstrated—Birth of the Goddess Durga, or the year, with the month Āswina.

THE early part of Astronomy among the Hindus, like that of all other nations, is involved in great obscurity. We can find no certain trace who the persons were that first began the science, nor the means employed by them for effecting their grand purpose; we are therefore obliged to pass over these as objects unattainable, and begin from the earliest known facts that offer themselves to our attention or investigation.

The most ancient part of the Hindu Astronomy, without doubt, is the formation of the Lunar Mansions; for, without a division of the heavens of some sort, or some fixed points to refer to, no observations on the positions of the Colures, or heavenly bodies, could be recorded with any degree of accuracy. History, and the poets, are perfectly silent as to the names of the first Astronomers, or the king in whose reign the science first began. All that we are informed is, that in the first part (quarter) of the *Tretā Yuga,* the daughters of Daksha were born, and that of these he gave twenty-seven to the Moon —that is to say, laying aside all allegory and poetic language, the twenty-seven Lunar Asterisms were formed in the first part of the *Tretā Yuga.* The *Tretā Yuga* began in the year 1528 B.C. and lasted about 627 years, the fourth part of which is 156¾ years; therefore the Lunar Asterisms must have been formed between the years 1528 and 1371 B.C. which might be considered as sufficiently near the truth. But as we have other means to approximate still nearer, it is proper we should notice them here.

It appears, that at the time of forming the Lunar Mansions, one of them, "*Visākhā,*" received its name from the equinoxial Colure cutting it in the middle, and thereby bisecting it, or dividing it into two equal sections, or branches; whence the name *Visākhā.* The observation here alluded to is mentioned in the Veda, and other books, and by which the positions of the Colures were as follows:—

The vernal equinoxial point was in the beginning of *Kritikā.*

The summer solstice in 10° of *Aślesha.*

The autumnal equinox in the middle of *Rādhā,* thence called *Visākhā.*

And the winter solstice in 3° 20′ of *Dhanisht'ha.*— See Plate I.

Now, in order to ascertain the time when this observation was made, we must find the precession from the position of some of the fixed stars at the time. Thus the longitude of Cor Leonis in the Lunar Mansion *Maghā* is always 9°. The vernal equinoxial point was found by the observation to be in the beginning of *Kṛitikā;* and from the beginning of *Kṛitikā* to the beginning of *Maghā* is seven Lunar Mansions of 13° 20′ each,

and therefore equal to	93°	20′
Add longitude of Cor Leonis in Maghā	9	0
Their sum is the longitude of Cor Leonis from Aries	102	20
Longitude of Cor Leonis in A.D. 1750, was	146	21
Difference is the precession . . .	44	1

or, the quantity by which the equinoxes fell back in respect of the fixed stars since the time of the observation. Now to find the number of years corresponding to this precession, it must be observed, that as we go back into antiquity, the rate of precession diminishes about 2″, 27 for every century. If we assume that the observation was made 1450 B.C. then $\frac{1450+1750}{2}=1600$; from which subtracting 1450, we get A.D. 150 for the middle point. Now in the first century of the Christian era, the precession was 1° 23′ 6″ 4, to which if we add 2″, 27, we get 1° 23′ 8″ 67 for the mean precession; that is to say, the precession that corresponds to the second century of the Christian era, in which the middle point is found. Therefore, as 1° 23′ 8″ 65 is to 100 years, so 44° 1′ to 3176 years, from which subtracting 1750, we get 1426 B.C. for the time of the observation, and the formation of the Lunar Mansions, which sufficiently corroborates what is stated in the *Kālikā Purāṇa*.

If this, however, should not be deemed sufficient for determining the time of the formation of the Lunar Mansions, we have other observations to mention that will be found to be still more accurate, as they can be depended on to the very year; and these are of the planets. From the union of the daughters of Daksha with the Moon, the ancient Astronomers feigned the birth of four of the planets; that is to say, Mercury from *Rohinī;* hence he is called *Rohineya,* after his mother. *Maghā* brought forth the beautiful planet Venus; hence one of the names of that planet is *Maghābhu.* The Lunar Mansion *Āshād'hā* brought forth the martial planet Mars, who was thence called *Āshād'hābhava;* and *Purvaphalgunī* brought forth Jupiter, the largest of all the planets, and the tutor of the gods: hence he is called *Purvaphalgunībhava;* the Moon, the father, being present at the birth of each. The observations here alluded to are supposed to have been occultations of the planets by the Moon, in the respective Lunar Mansions from which they are named*: they refer us to the year 1424-5 B.C. and therefore corroborating the result of the observation on the Colures. See Plate I.

The planet Mercury and the Moon in *Rohinī,* 17th April, 1424 B. C.;
The planet Jupiter and the Moon in *P. Phalgunī,* 23d April, 1424;
The planet Mars and the Moon in *P. Āshād'hā,* 19th August, 1424;
The planet Venus and the Moon in *Maghā,* 19th August, 1425;
all within the space of about sixteen months: and

* They are supposed to be occultations, because they are not made in the time of a single revolution of the Moon, but take in the space of about sixteen months, from 19th August, 1425, to the 19th April, 1424, B.C.; and this idea of the observations being confined to occultations is supported by Saturn not being included, because that planet was then out of the Moon's course.

there is no other year, either before that period or since, in which they were so placed or situated. Saturn is not mentioned among these births, probably from his being situated out of the Moon's course; but was feigned to have been born afterwards from the shadow of the Earth, at the time of churning the ocean, or the war between the gods and the giants, which will be noticed in its proper place.

It appears, that at first the number of Lunar Mansions was twenty-eight of 12° 51′ 3-7ths each; but that number being found probably inconvenient in practice, on account of the fraction, they were reduced to twenty-seven of 13° 20′ each. The first Lunar Asterism in the division of twenty-eight was called *Mulā*; that is to say, the root, or origin. In the division of twenty-seven, the first Lunar Asterism was called *Jyesht'hā*; that is to say, the eldest, or first, and consequently of the same import as the former. They both began from the same fixed point in the heavens, which was reckoned 2° 25′, or thereabouts, short of *Antarcs*.

The following are the Lunar Asterisms in their order, as exhibited in Plate I.

Division of 28 Mansions.		Division of 27 Mansions.	
1 *Mulā,*	15 *Mṛigasiras,*	1 *Jyesht'hā,*	15 *Mṛigasiras,*
2 *P. Āshād'hā,*	16 *Ārdrā,*	2 *Niriti,*	16 *Ārdrā,*
5 *U. Āshād'hā,*	17 *Punarvasu,*	3 *P. Āshād'hā,*	17 *Punarvasu,*
4 *Abhijit,**	18 *Pushyā,*	4 *U. Āshād'ha,*	18 *Pushyā,*
5 *Śravanā,*	19 *Aśleshā,*	5 *Śravanā,*	19 *Aśleshā,*
6 *Dhanisht'hā,*	20 *Maghā,*	6 *Dhanisht'hā,*	20 *Maghā,*
7 *Śatabhishā,*	21 *P. Phalgunī,*	7 *Śatabhishā,*	21 *P. Phalgunī,*
8 *P. Bhādrapadā,*	22 *U. Phalgunī,*	8 *P. Bhādrapadā,*	22 *U. Phalgunī,*
9 *U. Bhādrapadā,*	23 *Hastā,*	9 *U. Bhādrapadā,*	23 *Hastā,*
10 *Revatī,*	24 *Chitrā,*	10 *Revatī,*	24 *Chitrā,*
11 *Aswinī,*	25 *Swāti,*	11 *Aswinī,*	25 *Swāti,*
12 *Bharanī,*	26 *Rādhā,*	12 *Bharanī,*	26 *Rādhā,*
13 *Kṛitikā,*	27 *Anurādhā,*	13 *Kṛitikā,*	27 *Anurādhā.*
14 *Rohinī,*	28 *Indrā.*	14 *Rohinī,*	

* Abhijit, in the division of 28, is thrown out in the division of 27.

The next observations we meet with on record, bring us down to the winter solstice in the year 1181 B.C. when the sun and moon were in conjunction, and the Hindu Astronomers found that the Colures had fallen back 3° 20′ from their positions at the first observation: that is to say, the summer solstice was found in the middle of the Lunar Asterism *Aśleshā;* the autumnal equinox in 3° 20′ of *Viśākhā;* the winter solstice in the beginning of *Śravisht'hā;* and the vernal equinox in 10° of *Bharanī.*—See Plate II.

At this epoch, the Hindu Astronomy began to assume a more regular form—many improvements were made—the solar months were formed, and received their names—another set of Lunar Mansions was introduced, depending on the tropical revolution of the sun, corresponding in name and number with those of the zodiac, or fixed stars, and the six seasons of the year established on unalterable principles; which, with the months, depended also on the tropical revolution of the sun:—all of which are explained by the Plate, which shows their positions, as they then stood in respect of the fixed stars.

The outer circle contains the zodiacal Lunar Asterism, beginning with *Śravisht'hā,* as the first in the arrangement at that period, and numbered from 1 to 27. The next, the tropical Lunar Mansions, or those depending on the sun's revolutions in the tropics, coinciding with the astral ones at the time, and also numbered from 1 to 27.

The next circle contains the twelve months of the year in their order, beginning with *Māgha,* at the winter solstice. Next to these are the six seasons of the year, of two months each, and the first of which, *Śiśira,* begins at the winter solstice. The innermost circle of all represents the Serpent, which is the poetic emblem of both the year and the ecliptic. The head and tail meet at the winter solstice; and its middle corresponds with the middle of *Aśleshā*, which, in the tropical Lunar Mansions, was always the middle of the year, so long as it continued to be reckoned as commencing at the winter solstice. The following are the names of the Lunar Asterisms in their order, as numerically expressed in Plate II.

1 *Śravisht'hā,*	10 *Mṛigaśiras,*	19 *Chitrā,*
2 *Śatabhishā,*	11 *Ārdrā,*	20 *Swāti,*
3 *Purva Bhādrapadā,*	12 *Punarvasu,*	21 *Viśākhā,*
4 *Uttara Bhādrapadā,*	13 *Pushyā,*	22 *Anurādhā,*
5 *Revatī,*	14 *Aśleshā,*	23 *Jyesht'hā,*
6 *Aświnī,*	15 *Maghā,*	24 *Nirṛiti,*
7 *Bharanī,*	16 *P. Phalgunī,*	25 *P. Āshād'hā,*
8 *Kṛitikā,*	17 *U. Phalgunī,*	26 *U. Āshād'hā,*
9 *Rohiṇī,*	18 *Hastā,*	27 *Sravanā.*

The names of the moveable or tropical Lunar Mansions, always beginning from the winter solstice, are the same with the fixed or Astral Mansions; and therefore may sometimes cause an ambiguity, to be explained only by the nature of the subject. Thus when it is said, that the summer solstice is *always* in the middle of *Aśleshā*, we know immediately that the tropical or moveable *Aśleshā* is meant; just in the same manner as, if it was said that the summer solstice is *always* in the beginning of Cancer, we should know that the *sign* Cancer was meant, and not the *constellation* Cancer; because the solstices and Colures do not remain *always* in the same points with respect to the fixed stars.

It now remains to be explained, the principle on which the months were formed and named, and the time to which they refer. I have already observed, that the Lunar Mansions were fabled by the Hindu poets to have been married to the Moon, and that the first offspring of that poetic union were four of the planets. In like manner, the Hindu poets feign, that the twelve months sprung from the same union, each month deriving its name, in the form of a patronymic, from the Lunar Mansions in which the Moon was supposed to be full at the time.

Let us therefore, in the case before us, apply this principle. At the above epoch, 1181 B. C. the sun and moon were in conjunction at the winter solstice; and as the month began when the sun entered the signs, the first month therefore began at the winter solstice. Now to find the name of that month, the the moon would be full at about 14¾ days after the winter solstice, and would then be in the opposite part of the heavens to the sun. The sun would have advanced in 14¾ days about 14°½, and therefore would have entered the second Lunar Asterism, *Śatabhishā;* a line drawn from the point in which the sun is thus situated, through the centre, would fall into the Lunar Asterism *Maghā*, in which the moon was full, on the opposite side; and consequently, on the principle stated, the solar month was from thence called *Māgha,* in the form of a patronymic. At the next full, the moon would be in *Uttara Phalgunī,* and the solar month from thence called *Phālguna;* and on this principle all the months of the year were named; that is to say,

The month	*Māgha,*	from the Lunar Asterism		*Maghā,*	the 15th Mansion.
	Phālguna,	from	..	*U. Phalgunī,*	.. 17th.
	Chaitra,	from	..	*Chitrā,*	.. 19th.
	Vaiśakha,	from	..	*Viśakhā,*	.. 21st.
	Jyaishtha,	from	..	*Jyeshťha,*	.. 23rd.
	Āśhāra,	from	..	*P. Āshādťhā,*	.. 25th.
	Śrāvana,	from	..	*Sravanā,*	.. 27th.
	Bhādra,	from	..	*P. Bhādrapadā,*	.. 3rd.
	Āświna,	from	..	*Aświnī,*	.. 6th.
	Kārtika,	from	..	*Křitikā,*	.. 8th.
	Mārgaśirsha,	from	..	*Mrigaśiras*	.. 10th.
	Pausha,	from	..	*Pushyā*	.. 13th.

On the principle above stated, though the moon has been introduced by way of explanation, it is not at all necessary. All that is requisite to be understood is, that a line drawn from some part of the Lunar Mansion, through the centre, must fall into some part of that month to which it gives name, otherwise it does not answer the condition requisite. Hence, it is very easy to demonstrate the utmost possible antiquity of the time, when the months were, or could be, so named: for, there are certain limits beyond which the line cannot be drawn; and these are the termination of the Lunar Mansion, and the commencement of the solar month which determines the time; because, it points out the commencement of the solar month in respect of the fixed stars at the time. Thus, at the time of the above observations, the summer solstitial point was found in the middle of the Lunar Asterism *Áśleshā*, and the solar month *Śrāvana* then began; for, in the ancient Astronomy of the Hindus, that month always began at the summer solstice. Now the month *Śrāvana* derives its name from the Lunar Asterism *Śravanā* (the 27th), then in the opposite part of the heavens. (See the plate.) Let, therefore, a line be drawn from the solstitial point, or commencement of the month, cutting the centre, and it will fall into the very end of the Lunar Asterism *Sravanā*, from which it derives its name *Śrāvana;* which line is, therefore, at its utmost limit, as it cannot go farther without falling into a Mansion of a very different name. This position of the line, therefore, proves that the months received their names at the time of the above observations, and not before. For, if we wish to make it more ancient, let the solstitial point be supposed more advanced in respect of the fixed stars, say one, two, or three degrees, as at *a*, then a line drawn from the solstitial point, or commencement of the month *Śrāvana*, suppose at *a*, as in the dotted line, cutting the centre, cannot fall into any part of the Lunar Asterism *Śravanā*, from which it derives its name, but into *Śravisht'hā* (the 1st). Therefore, the name which it possesses, could never be given to it till the solstitial point, and commencement of the month, actually coincided with the middle of the Lunar Asterism *Aśleshā* (the 14th), being the same with the observation which refers us to the year 1181 B.C.; and this is the utmost antiquity of the formation and naming of the Hindu months, from which a very useful inference may be drawn, which is, that no Hindu writer, or book, that mentions the names of the Hindu months, can possibly be older than this period, let its pretensions to antiquity be ever so great.

Beside the observations on the positions of the Colures, there were others at the same time. Mercury was found at the winter solstice above mentioned, to be in and near the beginning of the Lunar Asterism *Śravisht'hā;* hence he is called *Śravisht'haja*, that is, born of the Lunar Asterism *Śravisht'hā*. This is, therefore, his second birth.

The commentator on the Astronomy of the *Rigveda*, also states that the planet Jupiter was in the first quarter of the Lunar Asterism *Dhanisht'hā*, that is, *Śravisht'hā*. Modern European tables make his place, as seen from the earth, more advanced by 13°; whence it appears that the position stated by the commentator was not from actual observation, but the effect of a computation made backwards, from about the 45th year of the Christian era, with the mean annual motion of Jupiter $1^s\ 0°\ 21'\ 9''$, 9, which quantity was employed before that time in settling the four ages hereafter mentioned in their proper place. This quantity is greater by 38″, 2, than our modern European tables make the mean annual motion of Jupiter; so that, if we divide 13° by 38″, 2, we get the 1225th year after 1181 B.C. or the year A. D. 45; from which point of time, the computation, being made backwards, would place Jupiter in 3° 20 of *Dhanisht'hā*, at the winter solstice 1181 B.C.

The astronomers of this period, after having formed and named the Hindu months, as above mentioned, framed a cycle of five years for civil and religious purposes, in which cycle they reckoned,

Sāvan days, or degrees	1800
Mean solar days	1830
Lunar days, or *Tithis* (of 30 to a Lunation)	1860
Lunar sideral days	2010
The number of Lunations, therefore, $\frac{1860}{30}$	62
The year, $\frac{1830}{5}$	366 days
The Lunation, $\frac{1830}{62}$	$29^{d.}\ 12^{h.}\ 23^{m.}\ 13^{s.}$

And the difference between the shortest and longest day was one hour and thirty-six minutes, which serves to point out the latitude of the place.

It is extremely probable that the above-mentioned observations were made in the time of Parasurāma, who, it is said, was a great encourager of astronomy. He lived upwards of 200 years before Rāma, whose time will be shown in the next Section. If we take the data given by Dr. Buchanan, in his Journey to *Malayala*, wherein he states, that they reckoned by cycles of 1000 years from Parasurāma, and that of the then current cycle, 976 years were expired in September 1800, there must be elapsed from the epoch of Parasurāma to A. D. 1800, 2976 years; from which taking 1800, we have 1176 B. C.

for the epoch of that prince, which differs but five years from the time of the above observation.

The years of the epoch of Parasūrāma are reckoned as beginning with the sign Virgo, or rather with the month *Āświna,* which was afterwards changed by the Chaldeans and Egyptians into the sign Virgo, at the time the constellations were formed; which will be noticed in its proper place. Some persons, perhaps, would think there was an error in commencing the year with the month *Āświna,* seeing that at the time of the observations, only five years before the epoch of Parasūrāma, the year began at the winter solstice with the month *Māgha.*

This is true; but it must be recollected, that the month of *Māgha* was not the only month with which the year could commence:—any month commencing at the same moment the sun entered a Lunar Asterism, or, in other words, when the month and Lunar Mansion began together, such month might begin the year. At the time of the above observations, there were three months, each of which began with a Lunar Mansion. Thus, the month *Māgha* began with the Lunar Asterism *Śravisht'hā:* the month *Jyaist'ha* began with *Mřigaśiras;* and *Āświna* began with the Lunar Asterism *Chitrā,* as may be seen by the Plate; and therefore the year might begin with either of them. The Lunar Mansions that begin with the months are called wives of the Sun, though they had already been all married to the Moon.

The commencement of the year with the month *Āświna,* of all others, was the most celebrated: Durgā, the year, personified in a female form, and goddess of nature, was then feigned to spring into existence. In the year 1181 B.C. the first of *Āświna* coincided with the ninth day of the moon;

and on that day her festival was celebrated with the utmost pomp and grandeur. In the year 945 B.C. some further observations were made, by which they determined, that in 247 years and one month, the solstices fell back 3° 20′ in respect of the fixed stars. In consequence of these observations, they threw back the epoch of the commencement of the year with *Āświna*, in 1181, to the year 1192 B.C., in which year the commencement of *Āświna* fell on the sixth day of the moon; and the festival of Durgā was ever after made to commence with the sixth lunar day of *Āświna*, and to continue down to the ninth inclusive, by which means both epochs were included. Thus I have shown both the origin and antiquity of the grand festival of Durgā, which of all others is the most ancient and the most superb. This goddess, properly speaking, signifies the year: she is therefore the goddess of nature; she is the consort of S̄iva (the personification of time in the male form); she is the same as the Juno of the *Greeks* and *Romans*, and the Isis of the *Egyptians;* she is Ceres, Proserpine, and in fact the same as all the goddesses; and their names are applicable to her.

SECTION II.

FROM 961 TO 698 B.C.

Rāma—*time of his birth determined from Astronomical Data—Date of the Rāmāyana determined—Churning of the Ocean, or War between the Gods and the Giants, description of it—Time of it determined from Astronomical and other Data—Birth of* Saturn—*The month* Kartika *made the commencement of the year—The rate of precession determined—A Lunisolar period discovered, on which were founded the changes to take place in the commencement of the year—A Table of eight periods of the same, and the year of their respective commencement, &c.*

In the last Section I gave the epoch of Paraśurāma, 1176 B.C. I shall now proceed to that of Rāma, the son of Daśaratha, and who is believed, or feigned by the modern *Hindus*, to have been one of the incarnations of the Deity. The epoch of this prince is considered the most famous in *Hindu* history, and perhaps deservedly so; for in his time, and that of his father, astronomy is said to have been cultivated with much attention; and it is supposed that the astronomical tables for calculating the places of the planets, were framed by means of the observations then made. It is, therefore, highly important that we should determine the time accurately, which fortunately we are enabled to do from astronomical data.

According to the *Rāmāyana* called Vālmika's, five of the planets were in their houses of exaltation, as the astrologers term it, at the birth of Rāma: that is to say, the sun was in Aries[1], the moon in Cancer, Venus in Pisces[2], Jupiter in Cancer[3], Mars in Capricorn[4], and Saturn in Libra[5], on the 9th lunar day of *Chaitra*.

The positions of the planets here given, I strongly suspect, are the result of modern computation, and not from actual observation: for the signs of the ecliptic, at least by these names, were totally unknown in the time of Rāma; and were not introduced into India, I believe, until the second or third century of the Christian era. However, be this as it may, the situations assigned to the planets, whether from computation or otherwise, point out to us, that Rāma was born on the 6th of April, 961 B.C.; at which time they were in the following positions:

	By Lalande's Tables.				By the *Rāmāyana.*
Sun	0s.	6°	11′	23″	In *Aries.*
Moon	3	12	13	54	*Cancer.*
Venus	11	1	0	0	*Pisces.*
Mars	10	2	47	0	*Capricorn.*
Jupiter	4	6	24	13	*Cancer.*
Saturn	6	8	27	0	*Libra.*

In which Jupiter is only 6° 24′ 13 beyond the limit, and Mars 2° 47′.

When Rāma attained the age of manhood, his father Daśaratha, in consequence of certain positions of the planets, approaching to a conjunction, supposed to portend evil, wished to share the government with him. Daśaratha says: "My star, O Rāma, is crowded with portentous planets; the sun, Mars, and the moon's ascending node" (*Rāhu*), &c. *Rāmāyana*, B. i. s. 3. v. 16.—"To-day the moon rose in *Punarvasu:* the Astronomers announce her entering *Pushyā* to-morrow: be thou installed in *Pushyā.*" v. 19.—"The sun's ingress into *Pushyā* being now come, the *Lagna* of *Karkaṫa* (Cancer), in which Rāma was born, having begun to ascend above the horizon," sec. 13. v. 3.—"The moon forbore to shine; the sun disappeared while it was day; a cloud of locusts, Mars, Jupiter, and the other planets inauspicious, approaching," sec. 33. v. 9, 10.

The facts pointed out here, show that there was an eclipse of the sun at or near the beginning of Cancer, at the moon's ascending node (*Rāhu* being present); and that the planets were not far distant from each other. These circumstances, therefore, point out the time to have been the second of July, in the year 940 B.C.; so that Rāma was then one-and-twenty years old. The following were the positions of the sun, moon, and planets at that time:—

Sun	2s	29°	5′	34″
Moon	2	29	5	34
—— Node ascending	3	9		
Mercury's geocentric L.	2	16		
Venus's do.	3	8		
Mars's do.	2	4		
Jupiter's do.	2	15		
Saturn's do.	3	0		

It appears from what is above said, that the beginning of *Pushyā*, and that of *Cancer*, were supposed to coincide; because it says that both the sun and moon entered *Pushyā:* now the fact is, that in the time of Rāma, no part of *Pushyā* coincided with *Cancer*. We are therefore led to this important conclusion, that the beginning of *Cancer* and that of *Pushyā* coincided when the author of the *Rāmāyana* wrote that work, and that he therefore concluded, though erroneously, that they were so in the time of Rāma. Now this gives us a clue to ascertain the date of the *Rāmāyana*. In the time of Rāma, the beginning of *Cancer*, or, which is the same thing, the beginning of the month *Srāvana*, coincided with 3° 20′ of the Lunar Asterism *Aśleshā;* and from thence to the beginning of *Pushyā*, is exactly 16° 40′. Now the beginning of *Cancer* must fall back 16° 40′ in respect of the fixed stars, before it could coincide with that of *Pushyā*—the precession was found equal to 3° 20′ in 247 years and one month—therefore

16° 40′ = 1235$^{yrs.}$ 5$^{m.}$ from which subtract 940, and we get A. D. 295, the time when the beginning of *Cancer* and that of *Pushyā* coincided, and consequently the period when the *Rāmāyana* was written. In thus giving the age of the *Rāmāyana* of Valmika, as it is called, I do not mean to say that the facts on which that romance was founded, in part, did not exist long before: on the contrary, my opinion is, that they did, and probably were to be then found in histories or oral traditions brought down to the time. The author of the *Rāmāyana* was more a poet than an astronomer, and being unacquainted with the precession, he fell into the mistake alluded to; for I do not suppose it was intentional, as that could answer no purpose. He made the like mistake, and from the same cause, in saying, that the moon at the birth of Rāma, on the 9th of the moon of *Chaitra,* was in the Lunar Asterism *Aditi,* or *Punarvasu,* which could not be the case.

I have now, I think, shown pretty clearly the epoch of Rāma from the positions of the planets at his birth, as well as at the time he was invited by his father to share the government with him. There is, however, another important circumstance that occurred in the time of Rāma, which we ought not to pass over, and which, while it also shows the time of that prince, points out a most material error in the chronology of the western world, in respect to certain points in history. The circumstance I allude to is what is generally called, or known by the name of, the *War between the Gods and the Giants.*

I am not aware that any person has ever attempted to determine the time of this extraordinary fiction, which is somewhat singular, as there are sufficient data for that purpose to be met with.

By what is stated in the *Rāmāyana*, it appears that Daśaratha, the father of Rāma, had two wives, and that he made a promise to one of them, at the time of churning the ocean, that her issue should succeed to the throne, in preference to the children of the other. That in consequence of this promise, when he wished that Rāma should be installed, he was opposed, and the compact he had unwarily entered into, brought forward as an argument against it; and as the promise of the king could not lawfully be broken, Rāma, his eldest son, and the real heir to the throne, was obliged to relinquish his right, and, in sorrow of heart, to betake himself to the wilderness, where he is said to have suffered great and many hardships.

The fact here stated, is sufficient to show the time of the churning of the ocean, otherwise called the *War between the Gods and the Giants*, that it must have taken place after the marriage of Daśaratha, nay, after the children were born; for it would be absurd to suppose that any such promise could well be made or exacted, before there was actual issue in being to benefit by it. However, be this as it may, we have other data to show, not only the year, but the very day and hour to which it refers.

Before we proceed, however, to the data, it will be proper, in this place, to give the Hindu description of the churning of the ocean, and the subsequent battle, as a principal part of the evidence will be found to emerge therefrom. On this occasion, I shall employ Mr. Wilkins's translation, as given by him from Book i. chap. 15. of the *Mahābhārata*, as being more full than what is to be found in the generality of the *Purānas*. It runs thus:—

"There is a fair and stately mountain, and its name is *Meru*, a most exalted mass of glory, reflecting the sunny rays from the splendid surface of its gilded horns. It is clothed in gold, and is the respected haunt of *Devas* (gods), and *Gandharvas* (celestial singers). It is inconceivable, and not to be encompassed by sinful man; and it is guarded by dreadful serpents. Many celestial medicinal plants adorn its sides; and it stands piercing the heavens with its aspiring summit—a mighty hill, inaccessible even by the human mind. It is adorned with trees and pleasant streams, and resoundeth with delightful songs of various birds. The *Sūras*, and all the glorious hosts of heaven, having ascended to the summit of this lofty mountain, sparkling with precious gems, and for eternal ages raised, were sitting in solemn synod, meditating the discovery of the *Amrita*, or water of immortality. The *Deva* Nārāyana* being also there, spoke unto Brahma† whilst the *Sūras*‡ were thus consulting together, and said, 'Let the ocean, as a pot of milk, be churned by the united labour of the *Sūras* and *Aśūras*§; and when the mighty waters have been stirred up, the *Amrita* shall be found. Let them collect together every medicinal herb, and every previous thing, and let them stir the ocean, and they shall discover the *Amrita*.'‖

"There is also another mighty mountain, whose name is *Mandār*, and its rocky summits are like towering clouds. It is clothed in a net of the entangled tendrils of the twining creeper, and resoundeth with the harmony of various birds. Innumerable

---

* Vishnu—time; 2nd person of the Hindu triad.
† Brahma—time; 1st person of the Hindu triad.
‡ *Sūras*; feigned deities, implying light.
§ *Aśūras*; feigned demons, implying the opposite to light, or darkness.
‖ *Amrita*; the fabled liquor of immortality.

savage beasts infest its borders; and it is the respected haunt of *Kinnaras* (celestial musicians), *Devas*, and *Aśūras* (celestial courtezans). It standeth eleven thousand yojans above the earth, and eleven thousand more below its surface. As the united bands of *Devas* were unable to remove this mountain, they went before Vishnu, who was sitting with Brahmā, and addressed them in these words: 'Exert, O masters, your superior wisdom to remove the mountain *Mandar*, and employ your utmost power for our good.' Vishnu and Brahmā having said, 'It shall be according to your wish,' he with the lotus eye directed the king of serpents to appear; and Ananta* rose, and was instructed in that work by Brahmā, and commanded by Nārāyana, to perform it. Then Ananta, by his power, took up that king of the mountains, together with all its forests, and every inhabitant thereof; and the *Suras* accompanied him into the presence of the Ocean, whom they addressed, saying, 'We will stir up thy waters to obtain the *Amṙita*.' And the lord of the waters replied, 'Let me also have a share, seeing I am to bear the violent agitations that will be caused by the whirling of the mountains.' Then the *Suras* and *Asuras* spoke unto *Kūrma-rāja*, the king of the tortoises, upon the strand of the ocean, and said, 'My lord is able to be the supporter of this mountain.' The tortoise replied, 'Be it so;' and it was placed upon his back.

"So the mountain being set upon the back of the tortoise, Indra began to whirl it about as it were a machine. The mountain *Mandar* served as a churn, and the serpent Vāsuki† for the rope: and thus in

* Ananta. The serpent Ananta implies time without end; also the year.
† The serpent Vāsuki, figuratively the year.

former days did the *Devas*, the *Asuras*, and the *Dānus* (or *Dānavas*, feigned giants), begin to stir up the waters of the ocean for the discovery of the *Amrita*.

"The mighty *Asuras* were employed on the *side of the serpent's head*, whilst all the *Suras* assembled *about his tail*. Ananta, that sovereign *Deva*, stood near Nārāyana.

"They now pull forth the serpent's head repeatedly, and as often let it go; whilst there issued from his mouth, thus violently drawing to and fro by the *Suras* and *Asuras*, a continual stream of fire, and smoke, and wind; which ascending in thick clouds replete with lightning, it began to rain down upon the heavenly bands, who were already fatigued with their labour; whilst a shower of flowers was shaken from the top of the mountain, covering the heads of all, both *Suras* and *Asuras*. In the mean time, the roaring of the ocean, whilst violently agitated with the whirling of the mountain *Mandar* by the *Suras* and *Asuras*, was like the bellowing of a mighty cloud. Thousands of the various productions of the waters were torn to pieces by the mountain, and confounded with the briny flood; and every specific being of the deep, and all the inhabitants of the great abyss which is below the earth, were annihilated; whilst, from the violent agitation of the mountain, the forest trees were dashed against each other, and precipitated from its utmost height, with all the birds thereon; from whose violent confrication a raging fire was produced, involving the whole mountain with smoke and flame, as with a dark blue cloud, and the lightning's vivid flash. The lion and the retreating elephant are overtaken by the devouring flames, and every vital being, and every specific thing, are consumed in the general conflagration. The raging flames thus spreading destruction on all sides, were at length quenched by a shower of cloud-borne water poured down by the immortal Indra.* And now a heterogeneous stream of the concocted juices of various trees and plants ran down into the briny flood.

" It was from this milk-like stream of juices, produced from those trees and plants, and a mixture of melted gold, that the *Suras* obtained their immortality.

" The waters of the ocean now being assimilated with those juices, were converted into milk; and from that milk a kind of butter was presently produced: when the heavenly bands went again into the presence of Brahmā, the grantee of boons, and addressed him, saying, ' Except Nārāyana, every other *Sura* and *Asura* is fatigued with his labour; and still the *Amṙita* doth not appear; wherefore the churning of the ocean is at a stand.' Then Brahmā said unto Nārāyana, ' Endue them with recruited strength; for thou art their support.' And Nārāyana answered and said, ' I will give fresh vigour to such as co-operate in the work. Let *Mandar* be whirled about, and the bed of the ocean be kept steady.'

" When they heard the words of Nārāyana, they all returned again to the work, and began to stir about with great force that butter of the ocean, when there presently arose from out of the troubled deep, first the *moon*, with a pleasing countenance, shining with ten thousand beams of gentle light; next followed Srī†, the goddess of fortune, whose seat is the

---

* Indra, regent of the skies — a personification of the sky.

† Srī, or Lakshmī, the lunisolar year; also the moon. — [Most of the names mentioned here, I believe, represent time.]

white lily of the waters; then Surā Devī*, the goddess of wine, and the white horse called *Uchīśrava*.† And after these, there was produced from the unctuous mass, the jewel *Kaustubha*‡, that glorious sparkling gem worn by Nārāyana on his breast: so *Pārijātaka*§, the tree of plenty; and *Surabhi*‖, the cow that granted every heart's desire.

"The moon, *Surā Devī*, the goddess Srī, and the horse as swift as thought, instantly marched away towards the *Devas*, keeping in the path of the sun.

"Then the *Deva* Dhanwantari¶, in human shape, came forth, holding in his hand a white vessel filled with the immortal juice *Amṙita*. When the *Asuras* beheld these wondrous things appear, they raised their tumultuous voices for the *Amṙita*; and each of them clamorously exclaimed, 'This of right is mine.'

"In the mean time, *Airāvata***, a mighty elephant, arose, now kept by the god of thunder; and as they continued to churn the ocean more than enough, that deadly poison issued from its bed, burning like a raging fire, whose dreadful fumes in a moment spread through the world, confounding the three regions of the universe with its mortal stench, until S̄iva††, at the word of Brahmā, swallowed the fatal drug to save mankind; which remaining in the throat of that sovereign *Deva* of magic form, from that time he hath

---

* Surā Devī, the goddess of wine, the consort of Bacchus — or the year.

† *Uchīśrava*, the white horse, probably the year; for time has long ears as well as wings.

‡ *Kaustubha*, the glorious sparkling gem, the sun.

§ *Pārijātaka*, the tree of plenty, is doubtless the year.

‖ *Surabhi*, the cow that grants every boon — the year.

¶ Dhanwantari, the year, or time, the best physician — the same as Esculapius.

** *Airāvata*. By this most probably is meant clouds, as productive of thunder and lightning, and appertaining to Indra, the sky, personified as the god of thunder.

†† S̄iva, time, the great deity of the *Hindus*.

been called *Nilkanta,* because his throat was stained blue.

"When the *Asuras* beheld this miraculous deed, they became desperate; and the *Amṙita* and the goddess Srī became the source of endless hatred.

"Then Nārāyana assumed the character and person of *Mohinī Māyā,* the power of enchantment, in a female form of wonderful beauty, and stood before the *Asuras,* whose minds being fascinated by her presence, and deprived of reason, they seized the *Amṙita,* and gave it unto her.

"The *Asuras* now clothe themselves in costly armour, and, seizing their various weapons, rush on together to attack the *Suras.* In the mean time, Nārāyana, in the female form, having obtained the *Amṙita* from the hands of their leader, the host of *Suras,* during the tumult and confusion of the *Asuras,* drank of the living water.

"And it so fell out, that whilst the *Suras* were quenching their thirst for immortality, Rāhu*, an *Asura,* assumed the form of a *Sura,* and began to drink also: and the water had but reached his throat, when *the sun and moon,* in friendship to the *Suras,* discovered the deceit; and instantly Nārāyana cut off his head, as he was drinking, with his splendid weapon *Chacra.*† And the gigantic head of the *Asura,* emblem of a mountain's summit, being thus separated from his body by the *Chacra's* edge, bounded into the heavens with a dreadful cry; whilst his ponderous trunk fell, cleaving the ground asunder, and shaking the whole earth unto its foundation, with

* Rāhu, the moon's ascending node personified.

† *Chacra,* the ecliptic, by which the dragon of old was feigned to be cut in two parts, *Rahu,* the head, or ascending node, and *Ketu,* the tail, or descending node.

all its islands, rocks, and forests. And from that time, the head of Rāhu resolved an eternal enmity, and continueth even unto this day, at times, to seize upon the sun and moon.*

"Now Nārāyana, having quitted the female figure he had assumed, began to disturb the *Asuras* with sundry celestial weapons; and from that instant a dreadful battle was commenced on the ocean's briny strand, between the *Asuras* and the *Suras*. Innumerable sharp and missile weapons were hurled, and thousands of piercing darts and battle-axes fell on all sides. The *Asuras* vomit blood from the wounds of the *Chacra*, and fall upon the ground, pierced by the sword, the spear, and the spiked club. Heads glittering with polished gold, and divided by the *Pattis* blade, drop incessantly; and mangled bodies, wallowing in their gore, lay, like fragments of mighty rocks sparkling with gems and precious ores. Millions of sighs and groans arise on every side; and the sun is overcast with blood, as they clash their arms, and wound each other with their dreadful instruments of destruction.

"Now the battle is fought with the iron-spiked club; and as they close with clenched fist, and the din of war ascendeth to the heavens, they cry: 'Pursue! Strike! Fell to the ground!' So that a horrid and tumultuous noise is heard on all sides.

"In the midst of this dreadful hurry and confusion of the fight, Nara † and Nārāyana entered the

---

* This eclipse at the war between the gods and the giants, so poetically described, is the first on record, that on the 2d July, 940 B.C. being the second. It is remarkable, that during the war between the gods and the giants in the west, which will be noticed in the next section, there was also an eclipse of the sun, which is the first known or mentioned in poetic story.

† Nara, the eternal, the same as Nārāyana or Vishnu.

field together. Nārāyana, beholding a celestial bow in the hand of Nara, it reminded him of the *Chacra*, the destroyer of the *Asuras*. The faithful weapon, by name *Sudarsana*, ready at the mind's call, flew down from heaven with direct and refulgent speed, beautiful, yet terrible to behold. And being arrived, glowing like the sacrificial flame, and spreading terror around, Nārāyana, with his right arm formed like the elephantine trunk, hurled forth the ponderous orb, the speedy messenger, and glorious ruin of hostile towns; who, raging like the final all-destroying fire, shot bounding with desolating force, killing thousands of the *Asuras* in his rapid flight, burning and involving like the lambent flame, and cutting down all that would oppose him. Anon he climbeth the heavens, and now again darteth into the field like a *Pisācha* to feast in blood.

"Now the dauntless *Asuras* strive with repeated strength to crush the *Suras* with rocks and mountains, which, hurled in vast numbers into the heavens, appear like scattered clouds, and fell, with all the trees thereon, in millions of fear-exciting torrents, striking violently against each other with a mighty noise; and in their fall, the earth, with all its fields and forests, is driven from its foundation. They thunder furiously at each other, as they roll along the field, and spend their strength in mutual conflict.

"Now Nara, seeing the *Suras* overwhelmed with fear, filled up the path to heaven with showers of golden-headed arrows, and split the mountain summits with its unerring shafts; and the *Asuras*, finding themselves again sore pressed by the *Suras*, precipitately flee. Some rush headlong into the briny waters of the ocean, and others hide themselves within the bowels of the earth.

"The rage of the glorious *Chacra, Sudarsana*, which for a while burnt like the oil-fed fire, now grew cool, and he retired into the heavens from whence he came. And the *Suras* having obtained the victory, the mountain *Mandar* was carried back to its former station with great respect; whilst the waters also retired, filling the firmament and the heavens with their dreadful roarings. The *Suras* guarded the *Amṙita* with great care, and rejoiced exceedingly because of their success; and Indra, with all his immortal bands, gave the water of life unto Nārāyana to keep it for their use."

In this highly coloured fiction, the Hindu poets have exerted all their abilities to give a most pompous description of a battle that never existed, the foundation of which shall presently be explained. By what is stated above, it will appear that the sun, moon, and Rāhu, or the moon's ascending node personified, were present, consequently an eclipse of the sun is thereby indicated at the ascending node. Moreover, the goddess Srī*, or Lakshmī, was then born, or produced from the sea. Therefore, in order to find the time, we refer to the Hindu calendar, where we find that her birth-day falls on the 30th lunar day of the moon of *Āświna;* so that the solar eclipse at the ascending node must have happened on that day; which circumstance alone would be sufficient to point out the day and year of the eclipse. But the goddess Lakshmī was born on a Thursday; hence that day is called *Laksmīwār;* and therefore

* The Venus Aphroditus of the Western mythologists, and emblematic of the lunisolar year: therefore she is called the goddess of increase, abundance, &c. She is the daughter of Durgā, and the Proserpine of the West; and, considered as time, she is the same with her mother. Metaphorically, she may sometimes represent the moon.

the eclipse must have been on Thursday.* From all these data, it is easy to determine the time, independent altogether of any knowledge of the time of Rāma. But beside these, there are others, if thought necessary. One is, that the planet *Saturn*, at the time of this eclipse, was supposed to be born from the earth's shadow; that is to say, that he was situated in that part of the heavens towards which the shadow of the earth projected: he was, in consequence, called the offspring of the shadow, or *Chāyāsuta*. Another is, that *Saturn* was born in the Lunar Asterism *Rohinī;* in consequence of which, they say, that any person born under that mansion, while *Saturn* is in it, is of the same nature with that planet, that is, of an evil disposition. The shadow of the earth at the time of the eclipse must therefore have pointed towards the Lunar Asterism *Rohinī*, in which Saturn was born. All these data are more than sufficient for our purpose. Proceeding, therefore, with the three first, we find that the eclipse we are in quest of fell on Thursday, the 25th October, in the year 945 B.C. at which time,

The *Sun's* longitude from *Aries* was about	6ˢ	22°	37′
The *Moon's* do. do	6	22	37
Sun from the *node* about	0	12	
Mercury's geocentric longitude about	6	13	
Venus's do. do.	7	11	
Mars's do. do.	6	11	
Jupiter's do. do.	8	26	
Saturn's do. do.	0	25	

From which it will appear, that all the planets, except *Saturn*, were on the same side of the heavens with the *sun* and *moon*. The *sun's* longitude at the time of the conjunction being 6ˢ 22° 37′, the point of the earth's shadow must be then directly opposite,

* The day was called Thursday, because Jupiter on that day conquered the Titans.

that is, in $0^s\ 22°\ 37'$; and as Saturn was in $0^s\ 25°$, the difference is something more than 2° from the line of the centre of the shadow; which difference, however, could be of no importance, as still *Saturn* would be considered in the earth's shadow, according to the notions of the ancients, who believed that the earth was much larger than the sun, nay, that it was supposed to be flat, and surrounded with various seas, until it reached the starry heavens. But *Saturn* was also said to be born in the Lunar Asterism *Rohinī;* which fact we must now ascertain, in order to prove the truth or falsehood of the assertion. For this purpose we make choice of the star *Cor Leonis,* whose longitude is 9° in the Lunar Asterism *Maghā.*

The longitude of this star, in A. D. 1750, was	4^s	26°	21′	12″
Subtract precession for 2694 years	1	7	40	42
Remain longitude in the year 945 B. C.	3	18	40	30
Subtract longitude of *Cor Leonis* in *Maghā*	0	9	0	0
Remain longitude of the beginning of *Maghā*	3	9	40	30
Subtract six Lunar Mansions, complete	2	20	0	0
Remain longitude of the beginning of *Rohinī*	0	19	40	30
Which taken from *Saturn's* longitude	0	25	0	0
Leaves Saturn's longitude in *Rohinī*	0	5	19	30

So that *Saturn* was actually near the middle of *Rohinī* at the time. *Rohinī* is the 9th Lunar Asterism, reckoning from *Dhanisht'hā,* as the first.

Thus I have at length, I think, not only confirmed the most famous epoch in *Hindu* history, that of Rāma, but also, in a most decisive manner, shown the real time of that most extraordinary of all fictions that ever was invented by human ingenuity, the War between the Gods and the Giants; which fiction, about two hundred years afterwards, was new-modelled and improved by Hesiod and others, and ultimately became the foundation of the religions of various nations of antiquity, on which more will be said in the next section.

I shall now proceed to give some explanation of the origin of the fiction, which in itself is nothing more nor less than a feigned war between light and darkness, and their imaginary offsprings. The eclipse took place on the 25th of October, at which time the longitude of the sun and moon was 6ˢ 22° 37′. Now the *Hindu* months always began at the moment the sun was supposed to enter the sign: therefore it was the 23d day of the month of *Kārtika* when the eclipse happened, reckoning back from this; and the first of the month will be found to have fallen on the end of the sixth day of the moon. This being the time of the autumnal equinox, it was found by observation that the Colures had fallen back in respect of the fixed stars, 3° 20′ since the former observations in 1181 B.C.; so that the cardinal points of the year, or the Colures, were now found in the following positions:—

The vernal equinoctial point in the middle of *Bharanī* . . . . . . the 7th.
The summer solstice in 3° 20′ of *Aśleshā* the 14th.
The autumnal equinox in the beginning of *Viśākā* . . . . the 21st.
And the winter solstice in 10° of *Śravanā* the 27th.

It is said, that at this time they formed the stars into various groupes or constellations, but of which we have now very little or no knowledge. It is also said, that the theogony was then formed or invented, and that the heavens were divided or shared between the gods, each having a certain portion assigned to him, which stands on Hindu record to this day.

Among these deities was Yama, the judge of the dead, the Pluto of the Western mythologists. To him the Lunar Asterism *Bharanī* (the 7th) was assigned as his house: so that from the positions of the Colures at the time above described, the equinoxial Colure passed through the very middle of it; and the solstitial Colures cutting *Aśleshā* in 3° 20′, and *Śravanā* in 10°, divided the heavens exactly in two equal portions; from which, I believe, the fiction of the heavens being divided between Jupiter and Pluto originally sprung. (See Plate III. in which the Mansions are numbered as in Plate II.) From these positions, it is evident, that the solstitial Colures, which divided the heavens, would also form the boundary between light and darkness; and as the serpent's head was always at the winter solstitial point, as I formerly explained, a line drawn from it, cutting the serpent in the middle, would place the head, or first half, in the darkened hemisphere, and the tail, or last half, in the enlightened hemisphere. Hence the poet, in his fiction, places the *Aśuras* at the head, and the *Śuras* at the tail, which we see was strictly true; for *Śura* means light, and comes from the root *Śura*, to shine, &c. and *Aśura* means the opposite, consequently darkness. Thus the morning star (which in ancient time was supposed to be the planet *Jupiter*), was called the guide of the gods, because it indicated the approach of day, or light; while, on the other hand, the evening star (*Venus*) was called the guide of the *Aśuras*, because it indicated the approach of darkness, or night. This short explanation will, I hope, be considered as sufficient: to enter more fully into the subject would require the whole of the Hindu theogony to be brought forward and explained, which would swell this essay to a much greater extent than what is absolutely necessary for the purpose intended.

It has been mentioned above, that the beginning of the month fell on the end of the sixth day of the moon, and that it was found that the Colures had fallen back from their former positions 3° 20′. This made the beginning of the month *Kārtika* coincide with that of the Lunar Asterism *Visākhā*, in consequence of which it was made the commencement of the year; and in order to make this circumstance still more remarkable, it was made the birth of Kartikeya, or the *Hindu god of war* (a personification of the year, beginning with the month *Kārtika*), naming him in the form of a patronymic, from the month with which the year began. Moreover, they established a festival in honour of him, which is marked in the Hindu calendar by the name of *Guha Shasti*. *Guha*, implying Kartikeya, and *Shasti*, the sixth day of the moon on which the year began. At the festival he is represented as riding on a peacock, indicating thereby that he is the head, and all the planets are stars in his train; whence, metaphorically, he is called the general of armies, which he is there supposed to lead. He is called by a variety of epithets, indicating his supposed exploits, qualities, &c. such as,

Shadānana—six-faced, in allusion to the six seasons of the year:

Dwādaśalochana—twelve-eyed, in allusion to the twelve months of the year:

Visākā—alluding to the commencement of the year with that Lunar Asterism.

It appears also, that the astronomers of this period (945 B.C.), among other things, had determined the rate of precession of the equinoxes, which they found to be 3° 20′ in 247 Hindu tropical years and one month; in consequence of which determination, they settled the commencement of the former period, and made it the first of Āswina in the year 1192 B.C.

which fell on the sixth day of the moon, as mentioned in the last section.

They found that in this period of 247 years and one month, in which the sun made 247 revolutions and one sign over, that the moon made 3303 revolutions and one sign over, and that there were 3056 lunations complete, and that the number of days in the whole period was 90,245½. Hence we get,

Length of the Tropical year	$365^{ds.}$	$5^{h.}$	$50^{m.}$	$10''\frac{98}{593}$
——— of the Hindu sideral year . .	365	6	9	$52\frac{3568}{6671}$
——— of a lunation .	29	12	44	3
Moon's tropical revolution	27	7	43	5

Mean annual precession 48″ 56661

Sun's mean motion for 365 days 11^{s} 29° 45′ 37″

Moon's mean motion for 365 days $13^{r.}$ 4 9 22 57

From the circumstance of the astronomical period above mentioned, containing one month over and above 247 years, it is obvious that it must begin and end with the same month of the year, and that the next succeeding period would begin with the month following, and thereby change the commencement of the year one month later each succeeding period: and, moreover, as there was a complete number of lunations (3056) in the period, it follows that the moon's age would be always the same at the commencement of each succeeding period. For instance, at the beginning of *Kārtika,* 945 B.C. the sun's longitude was 6^{s}, and the moon's was 8^{s} 12; hence it was the end of the sixth lunar day of 12° each: and this would be constantly the same at the beginning of each period, in succession, as may be seen by the following table of all the changes made in the commencement of the Hindu year, from 1192 B.C. down to A.D. 538, when the ancient method was entirely laid aside, and the present, or sideral astronomy introduced.

Periods	Began.	Months.	L. A. Coinciding.	Sun's Longit	Moon's Longit.	Calendar.	
1	1192 B.C.	1 *Āswina*	*Chitrā*	$5^s\ 0^0$	$7^s 12^0$	*Shasty Adikalpa*	1 Sept
2	945 ..	1 *Kārtika*	*Visākhā*	6 0	8 12	*Guha Shasti*	1 Oct.
3	698 ..	1 *Agrahāyanu*	*Jyest'hā*	7 0	9 12	*Mitra Saptami*	29 ..
4	451 ..	1 *Pausha*	*P. Āsād'ha*	8 0	10 12		27 Nov
5	204 ..	1 *Māgha*	*Sravanā*	9 0	11 12	*BhāscaraSaptamī*	25 Dec
6	44 A.D.	1 *Phālguna*	*Satabhishā*	10 0	0 12		23 Jan.
7	291 ..	1 *Chaitra*	*U.Bhādrapadā*	11 0	1 12		21 Feb.
8	538 ..	1 *Vaisākha*	*Aswinī*	0 0	2 12	*Jahnu Saptami*	22 Mar

In the above table, the first column contains the periods, the second the year before and after Christ, the third and fourth the *Hindu* month and Lunar Asterism with which the periods begin; the fifth and sixth, the sun and moon's longitudes at the commencement of each period; the seventh, the names and lunar days of commencement, according to the calendar; and the eighth, the corresponding day of the European months, according to common reckoning. Thus the first period began in the year 1192 B.C. on the first of the *Hindu* month *Āswina*, at the commencement of the Lunar Asterism *Chitrā*, the sun's longitude from Aries being then 5^s, and the moon's $7^s\ 12°$, corresponding to the 2d September, about sunrise, according to common reckoning in *India*. The commencement of each period falling at the end of the sixth lunar day, it might therefore begin either on the sixth or seventh; but the common practice afterwards being to commence on the following sunrise, the calendar states them all as beginning on the seventh, except the two first periods. The names *Mitra*, *Bhāscara*, and *Jahnu*, are those of the sun.

The third period began with the month *Mārgasirsha*, which name was changed into *Agrahāyana*, to express the circumstance of its commencing the year, and which it has ever since retained, though no longer beginning the year.

The precession of the equinoxes was reckoned from the commencement of the first period, in the year 1192 B.C., because in that year the moveable Lunar Mansions, or those depending on the sun's revolution in the tropics, coincided with the fixed or sideral ones of the same name; and the beginning of the solar month *Māgha,* which was always the instant of the winter solstice, and the commencement of the season *Śiśira,* coincided with the beginning of the Lunar Asterism *Dhanisht'ha,* otherwise called *Śravisht'ha.* All these, that is to say, the moveable Lunar Mansions, the solar months, Colures, and seasons, fell back in respect of their then positions with the fixed stars, at the rate of 3° 20′ in every astronomical period of 247 years and one month, or 48″,56661, annually.

SECTION III.

THE WAR BETWEEN THE GODS AND THE GIANTS IN THE WEST.

The War *between the* Gods *and the* Giants *in the West, described by* Hesiod *in his* Theogony—*The time of it determined from various Data—Its duration* 10 *years and five months—The Zodiacal Constellations and others then formed—The original idea of some of them appears to have been derived from the* Hindus—*The Tropical Signs named after the Zodiacal Constellations—The Months named after the Tropical Signs with which they then coincided—Their names compared—*Homer *and* Hesiod *not so ancient as generally supposed—The year represented by the Ancients under a great variety of names and personifications, as* Mercury, Hermes, Anubis, Budha, &c. &c.

In the foregoing section I have endeavoured to show the progress of Astronomy in *India* in the time of Rāma; we shall now take a view of the labours of the astronomers in other parts of the world, particularly in *Chaldea* and *Egypt*. Before, however, we can enter on this interesting subject, which will require particular attention, it will be necessary to insert here some parts of the fiction of the *War* between the *Gods* and the *Giants* in the *West*, the idea of which doubtless was borrowed from the one in the *East*, it being in like manner not only connected with, but serving essentially to point out the time of the formation of the constellations, to immortalize which seems to have been the object of the fiction.

Hesiod in his *Theogony* gives the following description:—

"When first the sire 'gainst Cottus, Briareus,
And Gyges, felt his moody anger chafe
Within him,—sore amazed with that their strength
Immeasurable, their aspect fierce, and bulk
Gigantic,—with a chain of iron force 825

He bound them down, and fixed their dwelling place
Beneath the spacious ground: beneath the ground
They dwelt in pain and durance in th' abyss,
There sitting where earth's utmost boundaries end. 830
Full long opprest with mighty grief of heart
They brooded o'er their woes: but then did Jove,
Saturnian, and those other deathless gods
Whom fair-hair'd Rhea bore to Saturn's love,
By counsel wise of earth, lead forth again 835
To light. For she successive all things told:
How with the giant brethren they should win
The glory bright of conquest.
Long they fought
With toil soul-harrowing; they the deities
Titanic and Saturnian; each to each 840
Opposed, in valour of promiscuous war.
From Othrys' lofty summit warr'd the host
Of glorious Titans: from Olympus, they
The band of gift-dispensing deities
Whom fair-hair'd Rhea bore to Saturn's love. 845
So wag'd they war soul-harrowing: each with each
Ten years and more the furious battle joined
Unintermitted: nor to either host
Was issue of stern strife nor end: alike
Did either stretch the limit of the war. 850

" But now when Jove had set before his powers
All things befitting: the repast of gods,
The nectar and ambrosia, in each breast
Kindled th' heroic spirit: and now all
The nectar and ambrosia sweet had shar'd, 855
When spake the father both of gods and men.
' Hear, ye illustrious race of earth and heav'n,
What now the soul within me prompts. Full long
Day after day in battle have we stood,
Oppos'd Titanic and Saturnian gods 860
For conquest and for empire: still do ye
In deadly combat with the Titans join'd,
Strength mighty and unconquerable hands
Display: remembering our benignant love
And tender mercies which ye prov'd, again 865
From restless agony of bondage risen,
So will'd our counsel, and from gloom to-day.'

" He spake; when answer'd Cottus the renown'd:
' O Jove august! not darkly hast thou said:
Nor know we not how excellent thou art 870
In wisdom; from a curse most horrible
Rescuing immortals. O imperial son
Of Saturn! by thy counsels have we ris'n

Again, 'from bitter bondage and the depth
Of darkness, all unhoping of relief, 875
Then with persisting spirit and device
Of prudent warfare, shall we still assert
Thy empire midst the rage of arms, and still
In hardy conflict brave the Titan foe.'

"He ceas'd. The gift-dispensing gods around 880
Heard, and in praise assented: nor till then
So burn'd each breast with ardour to destroy.
All in that day roused infinite the war,
Female and male: the Titan deities,
The gods from Saturn sprung, and those whom Jove
From subterraneous gloom releas'd to light, 885
Terrible, strong, of force enormous; burst
A hundred arms from all their shoulders huge:
From all their shoulders fifty heads up sprang
O'er limbs of sinewy mould. They then array'd
Against the Titans in fell combat stood, 890
And in their nervous grasp wielded aloft
Precipitous rocks. On th' other side alert
The Titan phalanx clos'd: then hands of strength
Join'd prowess, and display'd the works of war.
Tremendous then th' immeasurable sea 895
Roar'd; earth resounded: the wide heavens throughout
Groan'd shattering: from its base Olympus vast
Reel'd to the violence of gods: the shock
Of deep concussion rock'd the dark abyss 900
Remote of Tartarus: the shrilling din
Of hollow tramplings, and strong battle-strokes,
And measureless uproar of wild pursuit.
So they reciprocal their weapons hurl'd
Groan-scattering; and the shout of either host 905
Burst in exhorting ardour to the stars
Of heaven; with mighty war-cries either host
Encountering clos'd.

Nor longer then did Jove
Curb his full power; but instant in his soul
There grew dilated strength, and it was filled 910
With his omnipotence: at once he loos'd
His whole of might, and put forth all the god.
The vaulted sky, the mount Olympian, flashed
With his continual presence: for he pass'd
Incessant forth, and scattered fires on fires; 915
Hurl'd from his hardy grasp the lightnings flew
Reiterated swift; the whirling flash
Cast sacred splendour, and the thunderbolt
Fell: roar'd around the nurture-yielding earth
In conflagration, for on every side 920
The immensity of forests crackling blaz'd,

Yea, the broad earth burn'd red, the streams that mix
With ocean, and the deserts of the sea.
Round and around the Titan brood of earth
Roll'd the hot vapour on its fiery surge; 925
The liquid heat air's pure expanse divine
Suffus'd: the radiance keen of quivering flame
That shot from writhed lightnings, each dim orb,
Strong though they were, intolerable smote,
And scorched their blasted vision. Through the void
Of Erebus, the preternatural glare 930
Spread, mingling fire with darkness. But to see
With human eye, and hear with ear of man
Had been, as if midway the spacious heaven,
Hurling with earth shock'd—e'en as nether earth 935
Crash'd from the centre, and the wreck of heaven
Fell ruining from high. So vast the din,
When, gods encountering gods, the clang of arms
Commingled, and the tumult roar'd from heaven. 940
Shrill rush'd the hollow winds, and rous'd throughout
A shaking, and a gathering dark of dust;
The crush of thunders and the glare of flames,
The fiery darts of Jove: full in the midst
Of either host they swept the roaring sound 945
Of tempest, and of shouting: mingled rose
The din of dreadful battle. There stern strength
Put forth the proof of prowess, till the fight
Declin'd: but first in opposite array
Full long they stood, and bore the brunt of war. 950
Amid the foremost towering in the van
The war unsated Gyges, Briareus,
And Cottus bitterest conflict wag'd: for they
Successive thrice a hundred rocks in air
Hurl'd from their sinewy grasp: with missile storm 955
The Titan host o'ershadowing, them they drove
All haughty as they were, with hands of strength,
O'ercoming them beneath the expanse of earth,
And bound with galling chains; so far beneath
This earth as earth is distant from the sky: 960
So deep the space to darksome Tartarus."

Elton's Translation, p. 108 to 114.

Thus far Hesiod's description of the *War* between the *Gods* and the *Giants* in the West. He also gives a description of the battle between Jupiter and Typhæus, or the moon's ascending node personified; but as it contains no fact to point out the time, it would be useless to insert it here. There are many other descriptions of this *war*, differing from each other, as well as the places in which it was supposed to have commenced and ended, each poet endeavouring to transfer it to his own country. In some of these, we are told that the stars δ and γ of *Cancer*, called the *Aselli*, assisted Jupiter in his war with the *Giants*. This is a most material fact, because it serves to point out the time, independent of all other considerations. Hesiod notices the same thing, but in a very different manner: for, considering the inconsistency of two insignificant little asses assisting mighty Jove in his *war* with the *Titans*, he metamorphoses them with *Præsepe*, into three mighty *Giants*, by the names of Briareus, Gyges, and Cottus.

When the *War* between the *Gods* and the *Giants* was feigned to take place in *India*, 945 years B.C. the solstitial Colure cut the Lunar Asterism *Aśleshā* in 3° 20′, and the opposite one *Sravanā* in 10°; which Colure, therefore, divided the heavens in two equal portions, and formed the boundary between light and darkness at the moment of the autumnal equinox; the enlightened half belonging to the *Gods*, and the dark half to the *Aśuras*, called also, from their mother, *Dānavas*, and *Daityas*, the *Titans* of the west. At that period the stars δ, γ, and *Præsepe*, were in the dark half, or on the side of the *Titans*, as will be seen from their respective longitudes at the time, which were,

γ Cancri, 945 B. C.	2ˢ	26°	40′	14″
δ Cancri, ...	2	27	50	54
o Præsepe, (mean) .	2	26	19	52

Hence Hesiod, in alluding to the first *War* between the *Gods* and the *Giants*, that is to say, the one in *India*, poetically describes Cottus, Briareus, and Gyges, as bound with a chain of iron force, *beneath the spacious ground, in the abyss where earth's utmost boundaries end*, v. 822—830. The earth and sea were then supposed flat, and to extend to the starry heavens, in the same plane with the solstitial Colure, which was supposed to surround the earth as a wall of brass, leaving a passage to *Tartarus* in the beginning of *Cancer*, and another passage in the beginning of *Capricorn*—the former for the departed spirits to enter, the other to admit of their ascending to heaven, when their period of punishment terminated.

From the longitudes of the stars γ and δ of *Cañcer* in 945 B.C. it must be obvious that the *war* between Jupiter and the *Titans* must have been subsequent to that period; for these stars could not assist Jupiter until they were in the enlightened hemisphere, or at least in the boundary between light and darkness. Now the star δ is the nearest to that boundary, being distant from the beginning of *Cancer* only 2° 9′ 6″: the annual rate of precession at that period was 49″ 6; consequently the *war* in the *West* must have been at least 156 years later than the one in *India*. We know, however, from other circumstances, to be noticed hereafter, that it began about 33 years still later, that is to say, at the autumnal equinox in the year 756 B.C. and terminated at the era of Nabonassar. If the star γ had been employed in the calculation, the time would come out considerably later, that is to say, 241 years; to which if we add the former 156 years, and take the mean, we get 193 years; therefore 945—193 = 752 years, differing only four years from 756 B.C. To meet with extraordinary accuracy in ancient observations is not to be expected; therefore what is here exhibited must be considered as sufficiently exact to indicate the time. But over and above all this, there is another circumstance to be noticed, which is, that as the stars γ and

Præsepe have a more northern latitude than the star δ, they would appear in Egypt and other northern places to be in the horizon, or above it, when the star δ was in three signs, or the beginning of *Cancer*, at the time of the autumnal equinox at noon; so that in fact they would then be in the enlightened hemisphere, ready to co-operate with Jupiter against the *Titans*, according to the fiction of the poets. Hesiod therefore makes Cottus say: "O imperial son of Saturn! by thy counsels have we risen again, from bitter bondage and the depth of darkness, all unhoping of relief," v. 872—875. After the war was finished, Hesiod points out the station or place of Gyges, Cottus, and Briareus. "There the *Titanic gods* in murkiest gloom lie hidden; in a place of darkness where vast earth has end; from thence no open egress lies: Neptune's huge hand with brazen gates the mouth has closed; a wall environs every side. There Gyges, Cottus, high-soul'd Briareus dwell vigilant, the faithful sentinels of ægis-bearer Jove. Successive there the dusky earth and darksome *Tartarus*, the sterile ocean and the star-bright heaven arise and end, their source and boundary," v. 970—981. And again: "There night and day near passing, mutual greeting still exchange, alternate as they glide athwart the brazen threshold vast," v. 992—995. Here it is sufficiently plain that the *horizon* is meant, both east and west, as the points where the day begins and ends. At noon, at the autumnal equinox, the stars δ and γ, with *Præsepe*, would be in the western horizon; and at midnight, they would be in the eastern horizon.

In giving this explanation, my principal object was to point out, that the mighty giants Cottus, Gyges, and Briareus, of Hesiod, were no other than the stars δ and γ, called the *Aselli* by other poets, together with *Præsepe:* but whether they be considered so or not, is of no material consequence, as it is only to the positions of the stars we refer.

It may now be proper to mention the manner in which the *Aselli* assisted Jupiter. The stars δ and γ of *Cancer*, or the *Aselli*, being found in the enlightened hemisphere, and at or near the entrance into *Tartarus*, in the beginning of the sign *Cancer*, when the *Giants* attempted to ascend to the celestial regions that way, the *Aselli* set up such a braying noise as to dismay them completely from the attempt; they took to their heels, resolving to try the scaling of heaven in the opposide side; that is, where the gate or outlet was in the beginning of the sign *Capricorn:* but there they were equally disappointed; for Pan, that is to say, *Capricorn*, being placed there, on seeing them come near his gate, set up so great a noise, that even the *Gods* themselves were terrified with it; and the *Giants*, being *panic struck*, were glad to save their lives by a precipitate retreat.

This second *War* between the *Gods* and the *Giants* is noticed in some Hindu books, as having taken place in the next age after Rāma, in whose time I have shown the first took place.

Having thus far giving a short view of the *Wars* between the *Gods* and the *Giants*, and shown the times to which they respectively referred, we may now enter on the subject of the labours of the astronomers in the west, and other circumstances, during the period assigned to the last of these wars.

During the period of the first *War* between the *Gods* and the *Giants*, which, according to some accounts, lasted 10,748 days, the *Hindu* astronomers were employed in forming the stars into constellations, or groupes, under regular figures, and assigning to each of the Lunar Mansions its proper deity, drawn from the theogony, which is supposed to have been then, for that and other purposes, invented. The western astronomers, having, after the lapse of some years, received information of what was done in the east, conceived the idea of following the example, and of forming the stars into constellations also. The period employed for that purpose, in imitation of the eastern astronomers, was likewise termed the *War* between the *Gods* and the *Giants*, which is said to have lasted something more than ten years.

But though the astronomers of the west had thus far followed the notions of those of the east, and adopted their astronomy and rules, yet, in forming the constellations, they deviated from them both in their number and figures; and instead of dividing the zodiac into 27 or 28 constellations, as the Hindus had done, they adopted, in preference, as perhaps more convenient, a division of twelve, and in the forming of which they were guided by particular circumstances, some of which it may be proper to notice.

I have already mentioned, that the earth in ancient times was considered as flat, and that it was supposed to be surrounded with the sea on all sides, till it touched the starry heavens in the solstitial points, which were supposed to be always in the same plane therewith.

It is therefore obvious, that in these two opposite points, the constellations must be formed after such animals as were capable of existing in such a situation. The only animals that could be supposed capable of existing both in the sea and air, must be amphibious: the astronomers, therefore, placed at one point the constellation of a Crab, and in the opposite one that of a Sea-goat, both amphibious, and capable of living in or out of the sea. On the one side, the top of the head of the goat is about 8° above the water of the sea, and his chin nearly touching it; on the other side, the extremity of the head of the Crab is also about the same height above the sea; and on a level with the sea, the star δ in the breast of Cancer stands (see Plate IV.), to point out that the solstitial Colure then passed through it.

These two constellations being thus adjusted to their proper situations, all the rest would be easily settled, from the fables or fictions on which they were respectively founded. Some of them appear to have been derived from the *Hindus*. Thus the *Hindu* Durgā, and the Lion on which she is feigned to ride, were converted into the constellations *Virgo* and *Leo*, which may be proved from the position. Durgā, as I have already shown in the first section, was feigned to be born on the first of *Āswina;* which month, in the ancient astronomy of the *Hindus*, always coincided, from its beginning to end, with the sign of the *ecliptic* now called *Virgo*. The constellation Aries, and perhaps *Capricorn*, in part, I suspect to have been derived from the same source. *Aries*, or the *Ram*, is to be found in the ensign of Agni, who, according to the fictions of the *Hindus*, was feigned to ride that animal. The first half of *Kṙitikā*, over which Agni is considered as the presiding deity, corresponds with the end of the constellation *Aries;* and therefore the positions may be sufficiently near to warrant the conclusion drawn in this respect. But in respect to *Capricorn* some difficulty arises, from having neither position nor time of the year to guide us in the investigation. In the *Hindu* books, I mean the *Purānas,* time is personified under various shapes and names. Siva, which is the greatest of all the Hindu deities, is time. He is feigned, in these books, as marrying Durgā, the daughter of Daksha. Durgā, as already noticed, was a personification of the year, in a feminine form; and it would appear that Daksha, her father, was Cœlus, or the ecliptic, or, as some may have it, Atlas, as giving rise to all the Lunar Asterisms. Siva, his son-in-law, is said to have cut off his head, in consequence of not being invited to a feast given by Daksha; but repenting of the act, he restored him to life, and placed on him, in lieu of his own, the head of a goat. Here, to my apprehension, a change in the commencement of the sphere, or the year, is to be understood by the fiction, and that by such change, the year was made to commence from the winter solstice, which was in fact the case 1181 B.C.; and therefore I think it is from this fiction that the idea of the constellation *Capricorn* first arose. In like manner, some of the other constellations may have been derived from *Hindu* fiction. The constellation *Gemini* may, I think, have arisen from the story of the *Āswini Kumāras,* though we cannot deduce direct proofs at this time. But whether these conjectures be correct or otherwise, it is of no material consequence, as the object in view is the determination of the time when the constellations were framed, and not the various sources, from which the ideas respecting their formation may have originally sprung, which can only be of a secondary consideration.

The constellations of the *zodiac* having been all framed and named according to the animals or things intended to be represented under such figures, the astronomers then divided the *ecliptic* into 12 parts of 30 degrees each, commencing the division from the vernal equinoctial point To these divisions, which are called signs, they assigned the same names they had done to the constellations—the first sign, beginning at the vernal equinox, being called *Aries:* and as δ *Cancri* was then in 3^s, or the beginning of *Cancer,* the beginning of the sign *Aries,* or the equinoctial Colure, was of course three signs more to the west of that star.

The signs being named after the constellations, the next important step was to determine the year and the month, and to assign appropriate and significant names to the latter, corresponding either with the times of the year, or the signs of the ecliptic with which they then respectively coincided. If the year was precisely of the same length, as the space of time it would take the sun to pass, or run his annual course through the twelve signs, we should not be able to ascertain the time when they were first framed; because the months would always coincide with the same signs with which they began: but if there was any deviation, however small, by the constant accumulation of that deviation, the time would become known when they were first framed and settled. The *Egyptian* year, of all others, therefore, appears the best calculated for this purpose, and presents us with the requisite data. For, 1st, the *Egyptian* year contains only 365 days, and therefore falls back, with respect to the signs of the ecliptic, at the rate of 14′ 35″ 15 per annum; and, 2ndly, the names of the *Egyptian* months are also their names of the signs with which, at the time of their formation, they coincided, as will appear from the following table:

Egyptian Months.	Arabic Names.	Signs.
1 *Paophi*, or *Faopi*	*Foafi*	*Aries.*
2 *Athir*, *or Athyr*	*Thour*	*Taurus.*
3 *Chojac*	*Chayk*	*Gemini.*
4 *Tybi*	*Tab*	*Cancer.*
5 *Mechir*	*Mecheri*	*Leo.*
6 *Phamenoth*	*Famenoth*	*Virgo.*
7 *Parmuthi*, or *Farmuthi*	*Farmout*	*Equality.*
8 *Pachon*	*Buchomy*	*Scorpio.*
9 *Payni*	*Fayni*	*Sagittarius.*
10 *Epiphi*	*Hebhib*	*Capricorn.*
11 *Mesori*	*Mesour*	*Aquarius.*
12 *Thoth*	*Touhout*	*Pisces.*

The month *Thoth* begins the *Egyptian* year; but the names are here arranged according to the signs of the ecliptic, beginning with *Aries*, as the first.

Now the question is, when did the *Egyptian* months coincide with those signs of the ecliptic after which they are named?

When Julius Cæsar corrected the calendar, 44 B.C. the 30th of *Chojac* coincided with the 1st *January* at noon, at which time the

	s	°	′	″
Sun's mean longitude was, meridian of Paris	9	8	34	34
Subtract motion for 89 days .	2	27	43	21
Remain sun's mean longitude 1st *Paophi*, (4th Oct. 45 B.C.) .	6	10	51	13
Substract this from 12 signs, and we have	5	19	8	47

the quantity by which the month *Paophi* had retrograded, or fallen back, in respect of the sign Aries, from whence it derived its name. Now dividing this quantity by the annual variation or difference, we have $\frac{5-19-8 \cdot 47}{14 \quad 29}=702$ years; to which adding 45 B.C. we get the 747th year B.C. for the time when the months and signs of the same name coincided, and which corresponds with the era of Nabonassar. The same thing might have been determined much easier from the *Almagest;* for, at the era of Nabonassar, the

Sun's mean longitude on the first of *Thoth* at noon, was . .	11^s 0° 45
To which add 30 days' motion, .	0 29 34
We get the sun's mean longitude 1st *Paophi* at noon, ...	0 0 19

And as *Paophi* in the *Egyptian* language, meant *Aries*, both the sign and the month then coincided, or nearly so, there being only a difference of 19 minutes. The whole of this, therefore, proves beyond a doubt, that the constellations were formed and completely finished at the era of Nabonassar, and not before; and as this was the termination of the fictitious *War* between the *Gods* and the *Giants* in the West, neither Homer nor Hesiod could have written earlier than the year 746 B. C.

I am aware of the extreme difficulty of removing modern prejudices in favour of extraordinary antiquity, however unfounded the same may be. It may be said, that Hesiod's time must be known from the passage in his *Works and Days*, wherein he says:

> "When sixty days have circled since the sun
> Turned from his wintry tropic —— the star
> Arcturus, leaving ocean's sacred flood,
> First whole apparent makes his evening rise."

It may also be said, that the time of the *War* between the *Gods* and the *Giants* was considerably anterior to the era of Nabonassar; for that the names of certain kings are mentioned in history, at or near whose time this war is supposed to have taken place: and that from the positions of the Colures in the sphere of Chiron, as mentioned by Eudoxes and others, the time of the *Argonautic* expedition, and of the formation of the constellations, must be placed several centuries before the era of Nabonassar: moreover, that the *Egyptians* and *Chaldeans* received their astronomy from Abraham, and were therefore possessed of that science long before the year 746 B. C. and even immensely long before the *Hindus* had any knowledge of it, if we are to place any credit in the supposed antiquity of the zodiacs at Tentyra or Dendera, which men of science have made out to be about 15,000 years old.

I am perfectly aware of all these objections, likely to be made by persons who never gave themselves the trouble for a single moment to investigate the matter; but when all the circumstances and real facts are carefully and coolly examined without prejudice, the whole of such objections will, I am persuaded, fall to the ground, as they will be found to rest either on misconception, or on the stories of the priests and poets of antiquity, who framed them on purpose to answer their own views, as I shall now endeavour to point out.

First.—With respect to Hesiod, it is uncertain whether the rising of the star *Arcturus* meant the whole constellation, or the single star. It is also uncertain whether this rising was observed in Hesiod's own country, or even in Hesiod's time; particulars absolutely necessary towards making a just calculation. Moreover, the ancients were not very accurate in their observations of the solstices, which, in many instances, might differ a whole day or more from the truth, and thereby make a difference of upwards of 70 years in the result. Besides all these, it appears to have been the practice for the ancients to have copied from each other, without any regard to the latitudes of places, and to continue to express, or mention the times of the risings and settings of the stars that took place some centuries before them, as if actually occurring in their own times. Hesiod makes the *acronycal* rising of *Arcturus* sixty days after the winter solstice: Pliny, in particular, says the same. How can we then be certain but that Hesiod may have borrowed it also from some one before him? Herodotus, who was born B.C. 484, places Hesiod and Homer 400 years before his own time. But this, as well as what is said by others, and in the Chronicle of the Marbles, amount to nothing more than mere assertions without proof; nor can they stand a single moment opposed to the astronomical facts above mentioned, coupled with those yet to be noticed.

Secondly.—With respect to the *War* between the *Gods* and *Giants*, it is very easy to see the reason of its being placed some centuries farther back into antiquity than the real time of it. Were those who wrote first on this fictitious war, to state that it occurred between the years 756 and 746 B. C. which was the real time of it, they would be called impostors, because all others living at that period, would also be able to see it, as well as the writers themselves, if it really had taken place. To avoid, therefore, this dilemma, they found it necessary to throw it back into antiquity, out of the reach of the memory of any one then living, by which means they were enabled to establish it without fear of being contradicted, and to introduce into their religions, under the venerable appearance and sanction of *antiquity*, the idea of a place of punishment for the wicked, under the name of *Tartarus*.

Thirdly.—With respect to the positions of the Colures at the *Argonautic* expedition, we are informed by Eudoxes and others, that the solstitial

Colure cut the star δ *Cancri,* called the southern Asellus; consequently that star's longitude from the vernal equinoctial point, or beginning of the sign *Aries,* was then exactly three signs. In A. D. 1690, the longitude of the same star by the Brit. Catalogue was $4^s\ 4°\ 23'\ 40''$. The difference between the two longitudes was therefore $1^s\ 4°\ 23'\ 40''$. Reducing this to time, at the rate of 1° in 72 years, we get 2476 years, from which deducting 1690, we have 786 years before Christ, which might be considered as sufficiently near the truth. But to be still more accurate, let us take the middle point, or 1238 years before A. D. 1690, it will be A. D. 452. The precession, in the first century of the Christian era, was $1°\ 23'\ 6''\ 4$, and increased at the rate of about $2''\ 27$ each century: therefore, $2''.\ 27 \times 4 = 9''.08$, to which add $1°\ 23'\ 6''\ 4$, the sum is $1°\ 23'\ 15''\ 48$, the precession for the fifth century of the Christian era. Now as $1°\ 23'\ 15''\ 48$ is to 100 years, so $34°\ 40'$, the whole precession, to 2478 years; from which taking 1690, leaves 788 B. C. for the time of the feigned *Argonautic* expedition. But the periods of the formation of the constellations, of the *War* between the *Gods* and the *Giants,* and of the *Argonautic* expedition, being all the same, and having reference to the time between the *autumnal equinox* in the year 756, and the entrance of the sun into *Pisces* in 746 B.C., the above result differs from the truth only about 32 years; which is as near as can be expected from the observations of the ancients. The *Argonautic* expedition being, like the *War* between the *Gods* and the *Giants,* a mere fiction, it became, therefore, equally necessary to throw the supposed time of it back into antiquity, that the authenticity of the fabulous facts on which it was founded might not be controverted or overturned.

The position of the star δ *Cancri,* at the time of the *War* between the *Gods* and the *Giants,* and at the *Argonautic* expedition, being exactly the same, that is to say, in the beginning of the sign *Cancer,* shows, that the time must be the same, and that the one fact corroborates the other, in so far as time is concerned.

Lastly.—From all the facts thus exhibited, it appears sufficiently obvious, notwithstanding all that has been urged by writers on the supposed great antiquity of astronomy in *Egypt* and *Chaldea,* that, in reality, they had none before the era of Nabonassar, and that the whole of their pretensions were mere fictions, set up on purpose to conceal the source from whence they received it, as well as the time of its introduction; anxious to make themselves appear to be, not only the most ancient people on the face of the earth, but the first also in astronomy and other sciences. If, however, any other evidence be thought necessary to show that they were not so, we have the testimony of both Hipparchus and Ptolemy, who, after very diligent search, met with no observations made at *Babylon* before the time of Nabonassar. Epigenes speaks of *Babylonian* observations for the space of 720 years. Berosus allows them to have been made 480 years before his time, which carries them back to the year 746 B. C. We have, therefore, the most direct and positive proofs that could be given, that the *Chaldeans* and *Egyptians* had no astronomy, till they received it from *India* about the time above mentioned; and that it was then, and not before, that they were enabled to form the constellations, from the information and assistance they derived from the same source. It was not from Abraham they received their astronomy, as pretended, for the purpose of concealment, but from a *Brahman*, or the Brahmans: and though the word or name Abraham, might possibly arise by corruption, from pronouncing the name *Brahman*, yet I am disposed to think the corruption intentional.

But, should the whole of what is shown above, be deemed not sufficiently accurate, we have another fact of a still more decisive nature, which, while it serves to point out the time with the utmost precision, other circumstances connected with it, will be found to form the basis of an immense number of fictions, framed for the purpose of imposing on mankind. The fact that I allude to here is the *eclipse* of the *sun*, at the *ascending node*, which took place during the *War* between the *Gods* and the *Giants* in the West. Hesiod notices this *eclipse* under the fiction of the battle between Jupiter and Typhæus, or Typhon, which I mentioned in a former place: but he cautiously avoids saying any thing that could point out the time of it. He makes Jupiter to conquer him, who "down wide hell's abyss his victim hurl'd, in bitterness of soul." Others say, that the *sun* and *moon* chased Typhon till he was drowned in the *Thracian* sea, or, as others have it, till he was buried under Mount *Ætna:* but all these lead to no conclusion as to the time. The *Egyptians*, however, have not acted with the same caution; for, in alluding to this event, they say that Typhon put Osiris (the sun) into a chest, on the 17th day of *Athyr*, which therefore serves as a clue to discover the actual time of the *eclipse*. But, in order to limit our enquiries within proper or reasonable bounds, I have already shown, from other circumstances, that the *War* between the *Gods* and the *Giants* was from the *autumnal equinox* in the year 756 B. C. to the entrance of the *sun* into *Pisces* in the year 746 B.C. being ten years and five months: therefore, if this statement of the time be at all correct, the *eclipse* of the *sun* at the *ascending node* on the 17th day of the *Egyptian* month *Athyr*, must be found within that period. Beginning, therefore, our enquiry from the year 756 B.C. we find, at last, that the very eclipse we are in search of, fell on the 17th day of the month *Athyr*, in the year 751 B.C. being the sixth year of the *War* between the *Gods* and *Giants*, and therefore completely proving all that has been already stated. The *eclipse* took place in the afternoon, and *the sun set eclipsed in Egypt;* in consequence of which the *Egyptians* represented it by the figure of a *Hippopotamus*, (a *Typhonic* emblem,) receiving the *sun* in his jaws at setting in the western ocean. The *eclipse* must have been considerable, as the *sun* and *moon* at the time were very near the ascending node; but falling in the sacred period of the *War* between the *Gods* and the *Giants*, it could not be entered on record, like the subsequent *eclipses* after the era of Nabonassar, without exposing the whole imposition. The *time* was, therefore, kept a profound secret from all except the initiated and the priesthood, who had formed various festivals, on purpose to blind the populace, as well as for their own benefit.

Having thus far explained matters, we may now take a view of some of the other circumstances connected with this *eclipse*.

The longitudes of the *sun* and *moon* at the time of this *eclipse* was *Taurus*, 15° 46′, or thereabouts, and therefore in or very near the middle of the sign. The day on which the *eclipse* fell was made the birth of Hermes, the son of Osiris and Maia, one of the daughters of Atlas; or, which is the same thing, the birth of Mercury, the son of Jupiter and Maia; which day, according to the correction of the calendar by Julius Cæsar, in the year 44 B.C. was the fifteenth of the month now called *May;* so named, it is supposed, after the same Maia; and in consequence of which, the fifteenth of that month was always held sacred to Mercury in the *Julian* calendar.

Mercury, or Hermes, was considered by the ancients as emblematic of the year: hence we find the books of Hermes, as mentioned by some writers, as amounting to 360,00, or the number of degrees which the sun passes through in one year, multiplied by 100; or, as others have it, 365,24, being the number of days, &c. in the year, multiplied by 100: from which we gather, that the ancient tropical year of the *Egyptians* was 365 days 5 hours 45 minutes and 36 seconds, differing from the *Julian*, exactly one day in 100 years; which difference, in the time of Julius Cæsar, reckoning from the year of the *eclipse* 751 B. C. was upwards of 7 days: therefore, in order to make allowance for this difference, and to make his year correspond with the *Chaldean* reckoning, he was obliged to change the *Numaen* year into another form, making the 15th* of October, in Numa's year, to be the 1st of January in his own; so that the birth of Mercury fell on the fifteenth of *May*, the day on which the *eclipse* occurred.

MYTHOLOGICAL REMARKS.

From the circumstance of Hermes being considered the same as the year, he was called the great-

* See this matter more fully explained in my translations of two Hieroglyphic Calendars, (found at Dendera in Egypt,) at the end of this work.

est of chemists; for time reduces all things to their original states; compounds and decompounds bodies; converts earths into stones, and stones into earth; forming continual changes in every thing that exists: hence the science of chemistry was called the *Hermetic* art. The name Hermes we find also often united with others, forming thereby compounds generally expressive of the month, or sign, or other circumstance from which the year commenced: thus—Hermanubis, the year, commencing from the rising of the dogstar:—Hermaphrodite, the year, commencing from the first of April, from Venus presiding over that month, &c. Anubis was called sometimes the year, and marked the different seasons; but it seems the ancients had ascribed different meanings to Anubis, though all apparently expressive of certain points of time. He was called Anubis, or the watchful dog, that gave notice of the rising luminaries: he was, therefore, by some considered as the circle of the horizon; but this could only be at the moment of twilight. Anubis, considered as the year, had also the same offices assigned to him as Mercury had, such as being secretary to the gods, and recording all the events of a man's life, year by year, to appear for or against him at his death.

In different parts of *Egypt*, the year had different commencements. This was done, no doubt, to multiply festivals, and, consequently, emoluments to the priests. In some places, the year began on the entrance of the sun into the sign *Taurus;* at which time the Bull, *Apis,* was worshipped with great expense and ceremony. In other places, the year began at the winter solstice, and the Goat was worshipped; from which circumstance the people were called *Mendesians:* and in other places, the year began on the sun entering *Aries;* and that luminary was then worshipped under the name of Jupiter Ammon, and from whence the worshippers were called *Ammonians.*

In noticing the year under the names of Hermes, the son of Osiris and Maia, and Mercury, the son of Jupiter and Maia, we should not omit that of Buddha, the son of Māyā of the *Hindus.* This name is only a mere translation of Mercury, the son of Maia, Buddha and Mercury being the same: therefore Buddha, the son of Māyā of the *Hindus,* is nothing more or less than the year, which is even sufficiently shown, by other names assigned to him in the *Hindu* books. From this circumstance, it may be plainly perceived, that though the *West* had borrowed all its knowledge in astronomy from the *Hindus,* yet that the latter had, in their turn, borrowed from the *West.* This may be still further seen by the fiction of the *Hindu* Mars, usually called Kārtikeya, being born of the *Pleiades,* and thence called Bāhuleya, and, therefore, the same as Mercury, or Hermes. The greatest of all the *Hindu* gods is Siva, or *time,* and his image is generally accompanied with that of a Bull, to indicate the commencement of the year from the sign *Taurus;* and therefore, must have a reference to the ancient commencement of the year in the *West,* from the beginning of *Taurus,* or first of *May;* for there is no record or fact, from which we could draw a conclusion that the *Hindu* year ever began from *Taurus.* All these circumstances taken together, I think, sufficiently prove that a communication did exist at some former period, between the *East* and the *West,* though that period cannot be now ascertained. Woden, of the north of Europe, from his being supposed to have given his name to the *fourth* day of the week, as well as his presiding over *battles,* makes him the same as the Buddha and Mars (or Kărtikeya) of the *Hindus*, both of which meant the year.

Of all the gods of the ancients, none appear to have been personified into so many different shapes, or worshipped under so many different names, as the *year*, or *time*. It was made the foundation of the greatest part of their fictions, varying the personifications and worships, according to the purpose intended, the better to multiply the emoluments of the priests. In general, however, the personifications of *time*, or of the *year*, had their insignia, or characteristic symbols attached, by which they were known: this symbol was generally the Serpent, which of itself was known and understood to signify the *year*: sometimes wings were added, which had the same signification; for it is a common saying, that *time* hath wings. Hence the personification of the *year*, under the figure and name of Mercury, with his *caduceus* and wings,—also under the figure of Æsculapius, with a serpent in his hand, (which some, without just grounds, have taken to be a leech.) Among the *Romans*, the *first* of *January*, as the first of the year, was sacred to him. He was invoked and worshipped by the sick, or others for them; because, *time* and *patience*, are said to cure all diseases. Medusa, or rather Medusa's head, with serpents instead of hair, I have no doubt, also meant *time;* and the meaning of the fiction relating to her power of destruction, seems to have been this, that he who has seen her, must necessarily perish; that is to say, all who have been brought into existence must die. All the *Hindu* deities which have any relation to time, or the year, are in general accompanied with the symbolic serpent, as may be seen in sculptures and drawings of Siva, Krishna, and others. The figure of Siva is hardly ever seen without the Serpent. But it has also another distinguishing mark, which, as it serves to explain some of the customs of the *Egyptians*, ought not to be passed over. That is, the river *Ganges* is represented as springing from the head of Siva, the meaning of which is this: the celestial *Ganges*, or the *ecliptic*, begins at the commencement of the year, which is the head of *time*, and, therefore, represented as springing from the head of Siva. The *ecliptic*, in ancient *Hindu* books, is called the celestial *Ganges;* and this may serve to show, that in ancient times, the ecliptic was called a *river*. In *Egypt*, it was called the celestial *Nile;* and the gods and planets were therefore feigned to move along it in boats. At *Babylon*, it was called the *Euphrates;* and so on of other places.

It may be objected and said, that the serpent is an emblem of wisdom: true; but what is wisdom itself derived from but from years? A child just born cannot be said to have wisdom, nor can it be obtained but by the progress of time: hence, making the serpent as the emblem of wisdom, is only taking it at a second hand, that is, metaphorically derived from time. This may be seen in the *Hindu* name for Mercury, the son of Maia, or the year, which they call Buddha, the son of Māyā; thereby calling the year by the name wise, or wisdom, and consequently inferring wisdom as coming from years, the symbol of which is a serpent. We may also see, that Buddha, the son of Māyā, or the year, being the founder of a religious sect in India, is a mere fiction, and that we are rather to take it in a figurative sense; that is to say, that *time* has produced this religion, and that the person or persons, who may have first promulgated it, are now unknown. The epoch of Buddha is generally referred to the year 540 or 542 B. C. which probably may have been the period at which the religion of the *Bhuddhists* was first introduced into different parts of *India* and *China*. Time, in fact, may be considered as the author of every thing; for every thing is produced and perfected in *time*. *Time* is considered by the *Hindus* under three distinct points of view: first, as creating or producing; second, as preserving that which has been produced; and, third, as destroying ultimately that which has been preserved. These distinctions form the Hindu triad, under the names of Brahma, Vishnu, and Siva. All the offsprings of these, the feigned *Avatārs* of Vishnu, so called, as Buddha, Krishna, &c. are, therefore, in fact, only certain portions of *time*. Krishna has been considered by some as the *sun*: but this opinion is not correctly founded; for Vishnu being *time*, and as such always accompanied by serpents, the common symbol of *time*, no portion of him can be taken as the *sun*.* Krishna is considered by some, the same as the Apollo of the west: this, however,

* One of the mythological names of the *sun* is, I believe, *Garud'a*, the bird of Vishnu. Some have mistaken this for a real bird, as the eagle, &c.; but if his description had been particularly attended to, this, I think, could not happen. He is called by a variety of names, descriptive of what he is: such as, that he is brother of *Aruna*, or the dawn, and born after him; that he is of a beautiful plumage, and his body of a golden colour; and that he is the carrier of Vishnu; and that he is the destroyer of serpents, &c. All these appear to have a reference to the *sun*; for he rises after the dawn, is of a golden colour, and his rays are the beautiful plumage. He is the carrier of Vishnu, because the *sun* by his revolution carries *time* along with him. He destroys serpents, because the serpent is an emblem of the year; and therefore every past revolution of the *sun* is a year lost, or, metaphorically, a serpent destroyed; for *time* that is past cannot be recovered. The *Eagle* of Jupiter may also, perhaps, be considered in the same light.

can only be in respect of his qualities; for Krishna is a modern invention to serve a particular purpose. Apollo certainly is time personified, as well as Krishna, which may be proved by examining into his parentage, his inventions and discoveries. Jupiter and Latona are said to be his parents. Under the name of Orus, he was the offspring of Osiris and Isis in *Egypt*: in both cases, his father was probably the *sun* under a metaphorical name; and his mother, no doubt, though under different names, meant *time*: for Diodorus Siculus says, that Isis invented the practice of medicine, and taught the art to her son Orus; whence we must naturally conclude that Isis, and perhaps Latona, meant time taken generally; and that from the union of *time* with the *sun*, sprung Apollo, or the year. Hence, like Mercury, Hermes, Æsculapius, and other personifications of the year, he is said to have invented eloquence, music, medicine, and poetry, qualifications not applicable to the *sun*, but to *time*, as producing all things, according to the notions of the ancients, as well as of some of the moderns. Latona was the daughter of Saturn, or of *time*, and therefore must be considered as *time*, as well as Isis, who was also considered the same as Ceres, Juno, Luna, Terra, Minerva, Proserpine, Thetis, Cybele, Venus, Diana, Bellona, Hecate, Rhamnusia, and all the goddesses. From the whole of what is above shown, it is evident, that neither the *sun* nor *moon*, by whatever mythological names they may be called, can ever be considered as inventors of arts, letters, sciences, &c. except in metaphorical language; and then it is not the sun or moon that is to be understood, but their revolutions, or the time in which they are made, as years, lunations, &c.

A great deal more might be said in explanation of mythological subjects; but as I have already extended this article far beyond what I originally intended, I shall therefore now close it, and proceed to matters of more importance.

SECTION IV.

FROM 698 TO 204 B.C.

Commencement of the third Astronomical Period—The Precession then—The term Rishis *in* Maghā *explained—* Parāsara *and* Garga *cited—The Heliacal rising of Canopus in the time of* Parāsara*—The same computed—Positions of the Planets when* Garga *wrote—The time deduced—The real epoch of* Yudhisthira, 2526 *of the Kaliyuga, perverted by the Moderns—The fourth Astronomical Period.*

WE now come down to the third astronomical period, which, by the table at page 34, began in the year 698 B.C. on the first of the month *Margasirsha;* which name was now changed into that of *Agrahāyana*, from its commencing the year.

At the commencement of this period, the preces sion amounted to 6° 40 ; for all the months, seasons, Colures, and moveable, or tropical Lunar Mansions, depending thereon, were now found to have receded, or fallen back, in respect of the positions they had in 1192 B.C. by that quantity; and the beginning of the fixed Lunar Asterism *Maghā* was found in the middle of the tropical, or moveable *Maghā,* making a difference of 6° 40′ in 494 years and two months. To say that the beginning of *Maghā* fell into the middle of *Maghā,* would, to those who might be unacquainted with the real nature of the case, appear an inconsistency, though in fact there was none; for the same expression, or a similar one, we now use in saying that *Aries* is got into *Taurus:* meaning thereby, that the *constellation Aries* is got into the *sign Taurus,* which every one understands; and no doubt every one understood the meaning of the other at the time in the same way. But to do away every appearance of inconsistency, the astronomers of that period invented a term to answer the same purpose, and for showing the quantity of the precession. This was by assuming an imaginary line, or great circle, passing through the poles of the ecliptic and the beginning of the fixed *Maghā;* which circle was supposed to cut some of the stars in the Great Bear, which, by calculation, seems to have been the star β. The seven stars in the Great Bear being called the *Rishis,* the circle so assumed was called the line of the *Rishis ;* and being invariably fixed to the beginning of the Lunar Asterism *Maghā,* the precession would be noted by stating the degree, &c. of any moveable Lunar Mansion, cut by that fixed line, or circle, as an index. Thus, in the case above, where the beginning of the Lunar Asterism *Maghā* coincided with the middle of the moveable mansion *Maghā,* it would be expressed by saying, that the *Rishis* got into 6° 40′ of *Maghā,* that is, the line of the *Rishis* cut the moveable mansion *Maghā* in 6° 40′, thereby avoiding the supposed inconsistency that would arise in saying that one Lunar Mansion got into another, or that one part of the same mansion by name, got into a different one; which simple explanation brings us to the following one, respecting a passage in Parāsara, and which has greatly puzzled, or seemed to puzzle many of the moderns.

At the commencement of the first astronomical period, the fixed Lunar Asterisms, and the moveable ones of the same name, then coincided, (see Plate II.) and the positions of the Colures were therefore the same in both; that is to say, the winter solstitial point was in the beginning of both the fixed and moveable mansions *Śravisht'hā*, with which the beginning of the month *Maghā*, and that of the season of *Śiśira* also coincided. The summer solstitial point was in the middle of the fixed and moveable Lunar Mansions *Aśleshā*, with which point the beginning of the Solar month *Śrāvana*, and that of the season of *Varsha*, also coincided. The autumal equinoctial point was in 3° 20′ of the fixed and moveable Lunar Mansion *Viśākhā*, with which point the beginning of the solar month *Kārtika*, and the middle of the season *Sarada*, coincided. And the vernal equinoctial point was in 10° of the fixed and moveable Lunar Mansions *Bharanī*, with which point the beginning of the solar month *Vaisākha*, and the middle of the season of *Vasanta*, coincided. Now, as the moveable Lunar Mansions, depending on the sun's revolution to the tropics, with the Colures, months, and seasons, fell back together, and at the same rate of precession, (3° 20 in 247 years and one month,) it would naturally follow, as a matter of course, that the positions of the Colures in the moveable Lunar Mansions, would still remain invariably the same, and at all times, as they were in the year 1192 B.C. The following expression, therefore, of Parāsara, who, it will be shown, flourished about 575 years B.C. was perfectly correct and consistent. He says: "*When the sun having reached the end of* Sravanā *in the northern path, or half of* Ashleshā *in the southern, he still advances, it is a cause of great fear.*" And Garga, who was contemporary with Parāsara, and wrote his *Sanhita* in the year 548 B.C. says thus, to the same effect: "*When the sun returns, not having reached* Dhanishtā *in the northern solstice, or not having reached* Ashleshā *in the southern, then let a man feel great apprehension of danger.*"

In these two passages there is nothing inconsistent; for they are strictly conformable to the positions of the Colures, not only then but at all times, in the moveable Lunar Mansions, which alone are here meant, and not the fixed or sideral ones; there being no other mode in this case to distinguish between them, nor none necessary. The precession was well known in the time of Parāsara and Garga, and for upwards of 500 years before them; therefore, they could not be understood as speaking of the fixed Lunar Asterisms. If an European astronomer was now to say, "*If the sun, having reached the beginning of Cancer, still advances without returning great fear may be entertained*," no one could possibly misunderstand him—no one could say he meant the constellation Cancer: so it was the same in the time of Parāsara. These explanations are of the utmost importance, and should be particularly attended to, as they serve to show more clearly the impositions of the moderns, which will be noticed in their proper place.

There is another passage of Parāsara, respecting the *heliacal* rising of *Canopus* in his time, which it is proper to notice here. He states, that "*the star* Agastya *(or Canopus) rises heliacally when the sun enters the Lunar Asterism* Hastā, *and disappears or sets heliacally, when the sun is in* Rohinī."

As Parāsara was contemporary with Yudhishthira, and the latter ascended the throne in the year 575 B.C. we shall make the calculation for that year. The difference between the time of rising and setting of *Canopus*, points out the latitude of the place of observation to have been the same, or nearly the same as that of *Delhi*, which lies in latitude 28° 38 North.

The longtitude of *Canopus* in A. D. 1750 was..................	3s	11°	30′	40″				
Subtract precession for 2325 years. according to modern theory, ...	1	2	17	0				
Remain, ...	2	9	13	40				
Subtract variation in longitude of the star, by reason of the change in the ecliptic*, ...			26	30				
Remain true longitude of *Canopus* A. D. 575 B. C.					2s	8°	47′	10″
Latitude of *Canopus* A. D. 1750, ..		75	51	20				
Add variation for the change in the ecliptic*,..................			17	12				
True Latitude of *Canopus* 575 B. C.						76	8	32

From which data we now get the right ascension and declination of *Canopus* at that time by the following proportions :

		°	′	″	
1	As radius sine	90°	0′	0″	10.0000000
	Is to sine of the longitude of *Canopus*........	68	47	10	9.9695259
	So cotangent of the latitude................	76	8	32	9.3921570
	To cotangent of an ∠....................	72	2	55	9.3616829
	Subtract obliquity of the ecliptic 575 B. C...	23	46	46	
	Leaves..............................	3	16	9	
2	As radius sine	90	0	0	10.0000000
	To cosine of the longitude of *Canopus*	68	47	10	9.5585293
	So cosine of the latitude of *Canopus*	76	8	32	9.3793284
	To cosine of..........................	85	1	41	8.9378577
3	As radius sine	90	0	0	10.0000000
	To cosine............................	53	16	9	9.7767422
	So tangent of	85	1	41	11.0605043
	To tangent of the right ascension of *Canopus*..	81	43	25	10.8372465

Then for the declination:

		°	′	″	
4	As radius sine........................	90°	0′	0″	10.0000000
	To sine..............................	53	16	9	9.9038786
	So sine..............................	85	1	41	9.9983628
	To sine of the declination of *Canopus* S......	52	58	53	9.9022414

Having thus obtained the right ascension and declination of *Canopus* 575 years B. C. we are now

* These corrections might have been dispensed with, as the result would come out nearly the same without them.

prepared to proceed in the calculation of the *heliacal* rising of that star at the period given.

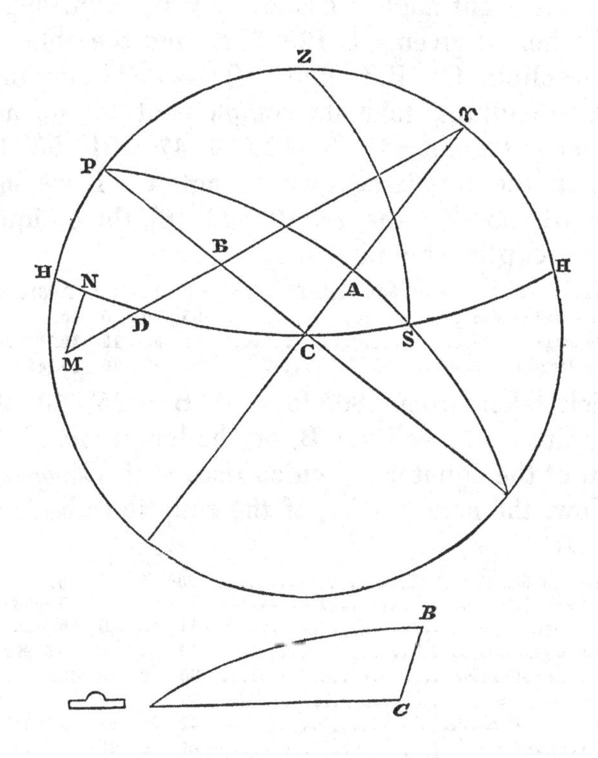

Let P be the pole of the equator.
PZ=90°—28° 38′=61° 28′=ACS
♈ C the equator.
♈ D the ecliptic.
S place of the star

5 As tangent of the colatitude PZ=ACS	61°	22′	0″	10.2628291
To tangent of the declination AS	52	58	53	10.1225946
So radius sine..........................	90	0	0	10.0000000
To sine of the ascensional difference AC	46	23	22	9.8597655
Add the right ascension, ...	81	43	25	
The sum is the oblique ascension ♈ C ...... the Equator rising with *Canopus*..........	128	6	47	or the point of

Now to find the point D of the ecliptic rising with S and C:

In the right-angled triangle ♈ CB, right-angled at C, there is given ♈ C 128° 6′ 47″ and the obliquity of the ecliptic C ♈ B, to find ♈ B:—♈ C being more than a quadrant, take its complement to the next equinoctial point=180°−128° 6′ 47″=51° 53′ 13″, then in the supplementary triangle C♎B we have C♎=51° 53′ 13″: the ∠♎=23° 46′ 46″, the obliquity of the ecliptic, to find B♎.

6	As cosine of the obliquity of the ecliptic = ∠♎	23°	46′	46″	=9.9614706
	Is to radius sine	90	0	0	10.
	So tangent C♎	51	53	13	10.1054247
	To tangent B♎	54	19	36	10.1439541

Which taken from 180° leaves ♈ B=125° 40′ 24″, the point of the ecliptic B, or the longitude of that point of the equator C, which rises with *Canopus*.

Now, the next portion of the ecliptic to be found is BD.

7	Say as radius sine	90°	0′	0″	10.
	To sine C♎	51	53	13	9.8958613
	So tangent ∠♎	23	46	46	9.6430682
	To tangent B C,	19	4	6	9.5389295
8	And as radius sine	90	0	0	10.
	To sine ∠♎	23	46	46	9.6055389
	So cosine C♎	51	53	13	9.7904366
	To cosine CB♎	75	35	21	9.3959755
9	And as radius sine	90	0	0	10.
	Is to cosine B C	19	4	46	9.9754623
	So tangent of the latitude of the place of observation,	28	38	0	9.7371709
	To cotangent of an ∠ which call *e*	62	42	26	9.7126332
	Which taken from C B♎ =75° 35′ 21″ leaves	12	52	55	which call *f*
10	As cosine ∠*f*	12	52	55	9.9889296
	Is to cosine ∠ *e*	62	42	26	9.6613749
	So tangent C B	19	4	46	9.5389246
	To tangent B D	9	14	26	9.2113699
	To which adding ♈ B	125	40	24	
	We have the longitude of the point D with S and C, or the *cosmical* point.	134	54	50	coascendent

The next portion of the ecliptic to be found is DM depressed 10° below the horizon.

11	As radius sine	90° 0′	0″	10.
	To cosine B D	9 14	26	9.9943272
	So tangent ∠ f	10 52	55	9.3592642
	To cotangent C D B	77 16	47	9.3535914
12	As sine N D M = C D B........................	77 16	47	9.9892080
	Is to sine M N	10 0	0	9.2396702
	So radius sine	90 0	0	10.
	To sine DM	10 15	15	9.2504622

Now the different portions of the ecliptic thus found, being added together, we have ♈ B + B D + D M = ♈ M = 125° 40′ 24″ + 9° 14′ 26″ + 10° 15′ 15″ = 145° 10′ 5″ = the sun's longitude from the vernal equinoctial point, when the star *Canopus* rose *heliacally* at *Delhi* 575 years before the Christian era; and as the *sun*, at the time of the *heliacal* rising of *Canopus*, was, according to Parāsara, in the beginning of the Lunar Asterism *Hastā*, the next thing we have to ascertain is the longitude of the beginning of *Hastā*, reckoned from the beginning of Aries, or the vernal equinoctial point, in the year 575 B.C.

For this purpose we make choice of the star *Cor Leonis*, whose fixed longitude is 9° in the Lunar Asterism *Maghā*.

The longitude of this star in A. D. 1750 from the beginning of *Aries* was	4ˢ	26°	21′	12″
Subtract its longitude in the Lunar Asterism *Maghā*	0	9	0	0
Remain longitude of the beginning of *Maghā* in A. D. 1750, ...	4	17	21	12
From the beginning of *Maghā* to the beginning of *Hastā*, is just three Lunar Mansions of 13° 20′ each; therefore add	1	10	0	0
The sum is the longitude of the beginning of *Hastā* in A. D. 1750, ...	5	27	21	12
Subtract from this the precession for 2325 years, (see before,)	1	2	17	0
Leaves longitude of the beginning of *Hastā* 575 years B. C.	4	25	4	12
Sun's longitude at the time of the *heliacal* rising of *Canopus* 575 B. C.	4	25	10	5
The difference is only	0	0	5	53

Which I think is a sufficient proof of the accuracy of the observation of Parāsara on the *heliacal* rising of *Canopus*. If the place of observation was a few miles more to the southward of *Delhi*, which is generally supposed to have been the case, and at a place called *Hastina-pura*, the ancient seat of government in the time of Yudhisht hira, then the agreement of the observation with the result of calculation would be still more perfect.

We have now to notice Garga, an ancient astronomer that was contemporary with Parāsara and Yudhisht hira. He states the positions of the *sun*, *moon*, and planets, at the commencement of the year (1st *Agrahāyana*), at sunrise, as follows:

The *sun* he states to be then in *Anurādha*,
The *moon* . . . in *Rohinī*,
Jupiter . . . in *Pushyā*,
Mercury . . . in *Satabhishā*,
Mars . . . . in *Mulā*,
Venus . . . . in *Kṛitikā*.

These positions, therefore, refer us to the 29th of October, in the year 548 B.C. the time when Garga wrote his *Sanhita*.

The facts we have thus exhibited prove decidedly, and beyond the possibility of doubt, the time in which Parāsara, Yudhisht'hira, and Garga lived, they being contemporaries. I have been more particular on this head than perhaps was necessary; but my reason for it was, to show the falsehood and impositions of certain modern *Hindu* writers, who have, through sophistry and low cunning, endeavoured to destroy the epoch of Yudhisht hira (2526 of the *Kaliyuga* of the modern astronomers), and to throw back his time to a very remote antiquity; some placing him 2448 years before the Christian era, while others, more bold, even go so far as 3100 years before that period. The impositions here alluded to, and the methods employed, will be more fully explained when we come down to the times of Aryabhatta, Varaha Mihera, and others.

The next astronomical period (the 4th) began on 27th of November 451 B. C. when the *Hindu* year commenced with the month *Pausha,* and the Lunar Asterism *Purvāshādhā.* In this period I have met with no observations worth mentioning, except one near the close of it in 215 B. C. when it was found, that at the winter solstice, or beginning of the solar month of *Māgha,* the *sun* and *moon* were in conjunction on Sunday, at sunrise, in the beginning of the Lunar Asterism *Śravanā.* This observation was the foundation of the festival called the *Ardha Udaya,* which is still kept up with great pomp and ceremony, though, strictly speaking, it cannot now take place under all the same circumstances, on account of the changes that have taken place by the precession of the equinoxes.

SECTION V.

FROM 204 B.C. TO A.D. 538.

Commencement of the fifth Astronomical Period — Astronomy further improved — More accurate Tables formed, and Equations introduced — The Hindu History divided into four Periods, and the Commencement of each settled astronomically — Tables of the four Ages, and their respective Years of Commencement, with the Errors in the Tables then used.

We now come down to the beginning of the 5th astronomical period, or 25th December 204 B.C. when the *Hindu* year began with the month of *Māgha*, at the winter solstice, and in the first point of the Lunar Asterism *Śravanā:* this is marked in the calendar with the word *Makari Saptami*, denoting, that the sun entered Capricorn on the 7th of the moon. Sometimes it is marked *Bhāskara Saptami.* The precession of the equinoxes now amounted to one whole Lunar Mansion, or 13° 20′, reckoning from the year 1192 B.C. when it was nothing; so that now the beginning of the fixed Lunar Asterism *Maghā*, was in the beginning or first point of the *moveable* Lunar Mansion *Purvaphalguni:* or, in other words, the *Ṛishis* were in the beginning of *Purvaphalguni:* that is, the assumed fixed line already mentioned as cutting the beginning of the fixed Lunar Asterism *Maghā*, and the star β of the Great Bear, did at this time cut the beginning of the moveable Lunar Mansion *Purvaphalguni:* so that the moveable Lunar Mansions, the months, and the seasons, all had fallen back 13° 20 from the positions they were in, 1192 B. C. in respect of the fixed stars.

It appears that at, or about this period (204 B. C.), improvements were made in astronomy: new and more accurate tables of the planetary motions and positions were formed, and equations introduced. Beside these improvements, the *Hindu* history was divided into periods, for chronological purposes; which periods, in order that they might never be lost, or, if lost or disputed, might, with the assistance of a few data, be again recovered, were settled and fixed by astronomical computations in the following manner: —The years with which each period was to commence and end having been previously fixed on, the inventor then, by computation, determines the month, and moon's age, on the very day on which *Jupiter* is found to be in conjunction with the *sun*, in each of the years so fixed on; which being recorded in the calendar and other books, might at any time be referred to for clearing up any doubt, in case of necessity.

It was from these conjunctions of the *sun* with *Jupiter*, that the periods themselves were named Yugas, or conjunctions; and the order in which they were named was thus: —The first period immediately preceding the inventor, was called the first, or *Kali Yuga;* the second, or next, was called the *Dwāpar Yuga;* the third was called the *Tretā Yuga;* and the fourth, or furthest back from the author, was called the *Kritā Yuga*, and with which the creation began. The end of the first period, called the *Kali*, was fixed by a conjunction of the *sun*, *moon*, and *Jupiter*, in the beginning of *Cancer*, on the 26th June, 299 B. C. This was called the *Satya Yuga*, or true conjunction, and is the radical point from which the calculation proceeds.

Having thus far explained the principles on which the four ages of the ancient *Hindus* were founded and settled, I shall now exhibit them complete, with all their dates, in the following table:

Names and Order of the Four Ages.	Dates.	Moon's Age and Month.	Error in the Hindu Tables used.
Kritā, or fourth	19th April 2352 B.C.	3rd *Tithi* of *Vaisākha*,	About 21° 46′—
Tretā, or third	28th Oct. 1528 ..	9th .. of *Kārtika*,	.. 13 1—
Dwāpar, or second	15th Sept. 901 ..	28th .. of *Bhādra*,	.. 6 22—
Kali, or first	8th Feb. 540 ..	15th .. of *Māgha*,	.. 2 33—
——— ended	26th June 299 ..	1st .. of *Srāvana*.	.. 0 1+

The mean motion of Jupiter in the *Hindu* tables employed for calculating the conjunctions and settling the periods, appear to have been $1^s\,0°\,21'\,9''\,54'''$, or nearly so, which being too great by about 38″ would cause the error to increase continually the further we go back into antiquity, as exhibited in the last column, and from which a near conclusion can be drawn as to the time the tables were framed, from the decrease in the error. I fix them to the year 204 B. C. because it was then the commencement of the astronomical period, at which the astronomers would naturally correct their table by new observations. Moreover, it appears that the *Hindu* history, according to the above periods, so settled and adjusted, was brought down, either by the inventor or some other person, to the year 204 B. C. and there terminated.

It will naturally be observed, that the year of the *Hindu* creation, or beginning of the *Kritā*, corresponds exactly with the year of the Mosaic flood, which is a most remarkable circumstance, and points out the opinion of the Hindus at that period, (204

B. C.) in respect of the time of the creation. The year 2352 B. C. was a leap-year, and the 19th April fell on Sunday, eight days after the vernal equinox. Thus the periods of *Hindu* history stood in the year 204 B. C.; but in the first century of the Christian era, it appears that they again changed the time of the creation, and carried it back to the autumnal equinox 4225 B. C. The particular periods then employed, the astronomical method used for fixing them, and the probable cause of the change, will be explained when we come down to that period.

The *Hindus* commence the reigns of their kings with the *Tretā*, which according to the table, began on the 28th of October, in the year 1528 B. C. common reckoning. Rāma, whose birth we have already shown, from astronomical facts, to have been in the year 961 B.C. was the last prince that reigned in the *Tretā:* and from the first, named Ikswāku, down to Rāma, inclusive, there were about 56 reigns in the space or period of 627 years, which gives an average of about 11 and 1-6th years to a reign. The *Tretā* terminated in the year 901 B.C.; and as Rāma was born 961 B.C. he must at the close of the period have been 60 years of age, if then living. The next period, the *Dwāpar,* began in 901 B.C. and ended in 540 B. C. and therefore lasted about 361 years, in which space there were thirty reigns, giving an average of about 12 years to a reign. The next period, the *Kali,* began in 540 B.C.: therefore Yudhishthira, whose time I have shown to have been 575 B. C. was of course but 35 years before the beginning of the *Kali Yuga.* The *Kali* lasted from 540 B.C. to 299 B.C.; but the reigns of the kings for that particular space of time are not distinguished from those that reigned after the period terminated, for a reason that will be hereafter explained. This, however, cannot prevent us from discovering the real period to which the reigns extended, taking them at the average already found; which, taking the two periods *Tretā and Dwāpar* together, gives an average of about 12 years to a reign. Now the number of princes in the solar line, that reigned after the commencement of the *Kali*, before that time became extinct, was 28; and 28 multiplied by 12, gives us 336 years for the period they reigned, which, being reckoned from the year 540 B.C. when the *Kali* began, will bring them down to the year 204 B.C. the very year at which the astronomical period commenced, and when the periods of the four ages were invented, as above stated: but what is equally remarkable, is, that the solar line of princes, the lunar line of princes, and the line of Jarashanda, should all become then extinct at one and the same time, as if the history after this period, was discontinued from some particular cause. The duration of the *Kali* from 540 B.C. to 299 B.C. being 241 years, the number of reigns in that period, at 12 years to a reign, would be 20; and from the year 299 B.C. down to 204 B.C. would be the eight remaining reigns, when the whole terminated. We may, from these circumstances, plainly perceive, that Vyāsa, the son of Parāsara, who lived 540 years B.C. was not the author of the ancient *Hindu* history, much less of its division into the periods above given, though pretended so to have been. Vyāsa could have given a history only to his own time, if he gave any; which, however, is very much to be doubted, as we find many other assertions of the modern *Hindus*, not only totally void of truth, but of the slightest foundation.

We shall now proceed to the next astronomical period, or the sixth, which began on the 23rd January A.D. 44, when the *Hindu* year commenced with the month *Phālguna,* on the 7th day of the moon, and with the Lunar Asterism *Satabhishā.* At the commencement of this period, observations were made, and the positions of the planets for that epoch, together with their mean motions, corrected, where necessary.

Early in this period, that is to say, about the year A.D. 51, Christianity was preached in India by St. Thomas. This circumstance introduced new light into *India,* in respect of the history and opinions of the people of the west, concerning the time of the creation, in which the *Hindus* found they were far behind in point of antiquity; their account of the creation going back only to the year 2352 B.C. which was the year of the Mosaic flood, and therefore would be considered as a modern people in respect of the rest of the world. To avoid this imputation, and to make the world believe they were the most ancient people on the face of the earth, they resolved to change the time of the creation, and carry it back to the year 4225 B.C. thereby making it older than the Mosaic account; and making it appear, by means of false history written on purpose, that all men sprang from them. But to give the whole the appearance of reality, they divided anew the *Hindu* history into other periods, carrying the first of them back to the autumnal equinox in the year 4225 B.C.: these periods they called *Manwantaras,* or patriarchal periods, and fixed the dates of their respective commencement by the computed conjunctions of *Saturn* with the *sun,* in the same manner as those of the four ages already given, were fixed by the conjunctions of *Jupiter* and the *sun*. This, no doubt, was done with a view of making the world believe, that such conjunctions were noticed by the people who lived in the respective periods; and therefore, might be considered as the real genuine and indisputable periods of history founded on actual observations.

The following table contains these periods, with their respective dates of commencement, &c.

Patriarchal Periods or Manwantaras.	Dates.	Moon's Age.	Errors in the Tables used.
1st,	25th Oct. 4225 B.C.	9th *Tithi* of *Āswina*,	30° 58′ 42″—
2nd,	13th Nov. 3841 ..	12th .. of *Kārtika*,	28 12 17 —
3rd,	11th April 3358 ..	3rd .. of *Chāitra*,	24 43 14 —
4th,	29th Aug. 2877 ..	3rd .. of *Bhādra*,	21 14 38 —
5th,	25th March 2388 ..	30th .. of *Phālguna*,	17 42 55 —
6th,	23rd Dec. 2043 ..	11th .. of *Pausha*,	15 13 6 —
7th,	2nd July 1528 ..	10th .. of *Āshād'ha*,	11 30. 8 —
8th,	8th Jan. 1040 ..	7th .. of *Māgha*,	7 58 22 —
9th,	28th July 555 ..	23rd .. of *Srāvana*,	4 28 28 —
Ended,	23rd June 31 A.D.	15th .. of *Āshād'ha*,	0 14 34 —

The mean annual motion of *Saturn* was $0^s\ 22°\ 14'\ 2''\ 48'''$, and the error in the mean annual motion $=26''+$; therefore the year in which there would be no error in the position of *Saturn*, would be A.D. 64, shewing the time when this division of the *Hindu* history was invented.

The introduction of this division into the *Hindu* history, occasioned no derangements in the times of the reigns of the princes of *India* by the former division:—for Ikswāku, the first king who began his reign at the commencement of the *Tretā*, 1528 years B.C. was transferred to the beginning of the seventh *Manwantara* in the above, or 1528 B.C. which, therefore, is the same time.*

* Ikswaku, on being so transferred, was called the son of the seventh *Menu*, who was feigned to be the offspring of the sun, which shows his origin to be fictitious; and from this fiction arose the appellation of solar line, being applied to his posterity. The lunar line, on the other hand, was feigned to have sprung from Buddha, the son

This division of the *Hindu* history was, however, doomed to be superseded by another about the year A.D. 538, in which the creation was thrown back 1972947101 years before the Christian era, and the real *Hindu* history entirely changed, as will be noticed and explained in the second part, when the subject of the modern astronomy is introduced.

The next astronomical period began in the year A D. 291, when the month *Chāitra* began the year, with the Lunar Asterism *U. Bhādrapadā*, and the month *Srāvana* (which was the same as the sign of the ecliptic Cancer), began with the Lunar Asterism *Pushyā*, that is, the beginning of Cancer and the beginning of *Pushyā* then coincided. I mention this merely to show that the *Rāmāyana* called Vālmīka's, could not be older than A.D. 292; but it might be a century, or even two later, the limit of the period in which it was written being from A.D. 292 to A.D. 538.

I have not been able to ascertain with sufficient certainty, the time when the tropical signs were first introduced into India; but they were certainly in use when the author of the Rāmāyana wrote, though probably not long before: we do not find the slightest mention of them in the genuine works of Parasara or Garga, nor in fact in any real work of antiquity.

There being no observations in this period, at least none on record worth mentioning, we therefore come down to the end of it, in A.D. 538, when the year,

---

of Soma, or the moon, by the Lunar Asterism *Rohini*, which is, therefore, also fictitious. The birth of this Buddha I have shown, in the first section, to have been the 17th April 1424 B.C.; consequently the solar line is older than the lunar by 104 years. From these circumstances, I think it highly probable that the Arcadians were a colony from India that settled in a part of Greece in early times, to which they gave their name (which is Sanscrit, implying descendants of the sun); for they called themselves older than the moon.

according to the regular periods above given, began on the first of *Vaisakha,* which then coincided with the beginning of *Aswinī,* on the 7th day, or Tithi of the moon. This beginning of the year was designated by the name *Jahnu Saptamī* in the calendars and other books of the *Hindus,* as one of the names of the sun. This year, A.D. 538, terminates the ancient astronomy of the Hindus, and commences the new, or modern, to which we shall now proceed.

PART II.

THE MODERN ASTRONOMY

OF

THE HINDUS.

SECTION I.

Commencement of the eighth Astronomical Period, the beginning of the Modern Astronomy — The Brahmins *introduce new and enormous periods into their History — The means adopted on the occasion — The new periods explained — fixed by Astronomical computations, the nature of which is explained at length — The revolutions of the Planets determined, and adjusted to the system of years so introduced — Method of determining the antiquity of the system, supposing the same unknown — The same by a Table of Errors continually decreasing down to the Epoch — The positions of the Stars given in* Hindu *Tables explained with a Diagram — Table of the Lunar Asterisms — The names of the Signs Aries, Taurus, &c. introduced from the West, and still used to represent the signs as beginning from the vernal equinoctial point — Some of the impositions of modern Commentators and others noticed — The system intended as a blow against the Christians — The* Avatars *invented for the same purpose —* Krishna *the Avatar noticed — His nativity computed from the positions of the Planets at his birth.*

In the preceding pages, I have endeavoured to give a clear and concise view of the ancient astronomy of the Hindus, so far as the same was found to be connected with history, from the earliest dawn of its commencement, down to A. D. 538, which was the beginning of the eighth astronomical period.

This epoch is one of the greatest importance, not only in Hindu history, but also in astronomy, as it was now that means were adopted by the Brahmins for completely doing away their ancient history, and introducing the periods now in use; by which they threw back the creation to the immense distance of 1972947101 years before the Christian era; with a view, no doubt, to arrogate to themselves that they were the most ancient people on the face of the earth.

The various means or contrivances that were adopted for this purpose will now be explained.—In the first place, they made choice of a period of 4320000000 years, which they called the *Kalpa.** This period they divided and subdivided into lesser periods, which, the better to answer their purpose, they called by the same names as the periods of the two former divisions of the *Hindu* history were designated, in order that they might be conceived to be the same.

The *Kalpa*, or 4320000000 years was divided into 14 *Manwantaras*,† each consisting of 308448000 years, with the addition of 1728000 years to make up the *Kalpa*. The *Manwantara* they divided into 71 *Mahā Yugas*, or great ages of 4320000 years each, with the addition of 1728000 to make up the *Manwantara*. The *Mahā Yuga*, or great age, they divided into four others, viz. the *Kali* of 432000

---

* *Kalpa* implies form.

† The division of their history into *Manwantaras*, as formerly given, only consisted of nine, the first of which began in the year 4225 B. C. at the autumnal equinox, and the last terminated in the year A. D. 31; whereas they made their modern *Kalpa*, as above, to consist of fourteen *Manwantaras*, and therefore not the same number: this objection they foresaw, and, to obviate the force of it, added five nominal, or spurious ones to the former nine, to make their number the same with the modern ones, and, to give them a better appearance, inserted their pretended dates in the calendar and other books, as in the following Table:

The spurious *Manwantaras* added,

10. Date, 15th of the moon of *Ashād'ha*,
11. .. 15th of the moon of *Kārtika*,
12. .. 15th of the moon of *Phālguna*,
13. .. 15th of the moon of *Chaitra*,
14. .. 15th of the moon of *Jyest'ha*.

years, the *Dwāpar* of 864000 years, the *Treta* of 1296000 years, and the *Kritā* of 1728000 years; the four making up the number of years in the *Mahā Yuga*, or great age=4320000 years: thus giving to these periods, for the sake of imposition, the same names they had done in the former divisions of their history.

The *Kalpa*, being divided and subdivided into the periods above given, the next step was to fix the commencement of the *Kalpa* itself, and consequently the creation, which was assumed to have then taken place.

For this purpose, it was resolved to frame an astronomical system, in which the planetary motions were to commence with the *Kalpa*, and to make the computation of eclipses, and the positions of the planets, at all times, to depend on that circumstance; by which means an air of truth and reality would be given to the whole, in the same manner as if actual observations had been made at the beginning of the *Kalpa*, or creation. In framing such a system, they resolved to adopt the sideral sphere and year, in place of the tropical, which till then had been in use; so that the beginnings of the months and years would always, for the future, remain fixed to the same points, in respect of the fixed stars, in which they then stood; and be also the same at the beginning of the *Kalpa*, or creation. Matters being thus far settled, the next step was to ascertain, by computation, a point of time from which the calculation of the length of the year and the mean motions of the planets should proceed, in order to determine the number of revolutions of each in the *Kalpa*, preparatory to their application to astronomical purposes. The only point of time they could find to answer this purpose was the 18th of February, in the year 1612 of the Julian period; and this point they made the commencement of the *Kāli Yuga* of the twenty-eighth *Mahā Yuga,* of the seventh *Manwantara:* from which we are now enabled to shew the number of years then elapsed of the *Kalpa,* or in other words, from the creation, according to this new system, as follow :—

Period of years at the beginning, called a *Sandhi,* =......	1,728,000
Six *Manwantaras* complete, or 308448000 × 6	1,850,688,000
Twenty-seven *Mahā Yugas* of the 7th *Manwantara,* or 4320000 × 27 = ...	116,640,000
Kritā of the 28th *Mahā Yuga,*	1,728,000
Tretā of the same, ...	1,296,000
Dwapar of the same, ...	864,000
To the beginning of the *Kaḷi Yuga,* (or 18th February, 1612 J. P.) = ...	1,972,944,000

The point of time thus fixed on, was found by computation made backwards, which showed that the planets were then approximating to a mean conjunction in the beginning of the sideral sphere, commencing with the Lunar Asterism Aswinī; on which account it was made choice of as the point to proceed from: for, had the approximation of the planets been in any other part of the heavens, it would not have answered their purpose; because their object was to assume the sun, moon, and all the planets, to be then in a line of mean conjunction in the beginning of *Aswinī,* or the sideral sphere, in order that from that assumption, as if it had been an actual observation, they might determine the length of the year, and mean motions of the planets, sufficiently near the truth to answer their purpose: for, whatever errors there might be in such an assumption, the same being divided among the years elapsed when the system was framed, would appear so small as not to be worth notice.

For the better understanding of this, it will be proper to give here the positions of the planets at the point fixed on, viz. on the 18th February, in the year 1612 of the Julian period, at sunrise, on the meridian of *Lanka*, or, more properly speaking, the meridian of *Ujein*, where this system was invented, and which is about 75° 50′ E. of Greenwich.

MEAN PLACES OF THE PLANETS AT THE GIVEN TIME.

	European Sphere.										Hindu Sphere.			
Sun, ...	10ˢ	1°	15′	48″	+	1ˢ	28°	44′	12″	=	0ˢ	0°	0′	0″
Moon, ...	10	4	24	36	+	1	28	44	12	=	0	3	8	48
Mercury, ...	8	28	36	49	+	1	28	44	12	=	10	27	21	1
Venus, ...	11	4	8	39	+	1	28	44	12	=	1	2	52	51
Mars, ...	9	19	3	11	+	1	28	44	12	=	11	17	47	23
Jupiter, ...	10	18	5	9	+	1	28	44	12	=	0	16	49	21
Saturn, ...	9	10	2	28	+	1	28	44	12	=	11	8	46	40
Moon's apogee,	2	1	14	7	+	1	28	44	12	=	3	29	58	19
——Node supt.	7	5	22	16	—	1	28	44	12	=	5	6	38	4

The sun at the given moment is supposed to be just entering the first sign in the *Hindu* sphere; but its longitude at the same instant in the European sphere was 10ˢ 1° 15′ 48″, the difference 1ˢ 28° 44′ 12″ is the difference between the two spheres at the time; which being added to the longitudes of the planets in the tropical sphere, reduces their places to the *Hindu*; from which it may be easily seen that the planets were not in the position assumed, and that the errors in the assumption so made were,

In the Sun's place, ...	0°	0′	0″	
Moon's do.	3	8	48	—
Mercury's do.	32	38	59	+
Venus's do.	32	52	51	—
Mar's do.	12	13	37	+
Jupiter's do.	16	49	21	—
Saturn's do.	21	13	20	+

The marks or signs — + show, that the assumed position falls short of, or exceeds the real mean place, by the quantity to which they are annexed: thus the position assumed being 0^s, falls short of the moon's mean place at the time by $3° 8' 48''$, and exceeds the mean place of Mercury by $32° 38' 59''$, because Mercury was then only in $10^s 27° 21' 1''$ instead of $0^s 0° 0' 0''$, the assumed position.

From the circumstance here stated, it must be self-evident, that in deriving the mean annual motions of the planets from the assumed position at the given time, the mean motions of the moon, Venus and Jupiter, must come out greater, and those of Mercury, Mars, and Saturn, less, than the truth, —and that the errors in such mean annual motions would, if nothing else operated to the contrary, be in proportion to the errors above exhibited in the position assumed.

Having thus shown the principal cause of the difference between the Europeans and modern *Hindus* in respect of the *quantities* of the mean annual motions of the planets, we may now proceed to determine the mean motions themselves, preparatory to our showing the manner in which the astronomical system was formed and connected with the system of years already mentioned.

I have already stated, that the ancient astronomy of the *Hindus* terminated in March, A.D. 538, at which period the new system was introduced. In this year the vernal equinoctial point, the beginning of the Lunar Asterism *Aswini*, and the beginning of the month Vaisākha, were supposed to coincide, (See Plate VI.) which point was, therefore, made the commencement of the year in this new system; so that the sun was supposed to enter into the sign *Aries* of the tropical sphere, and into the first sign of the Hindu sphere at the same moment of time. Now the instant of the mean vernal equinox in that year, was the 21st March, about six in the morning, on the meridian of Ujein, and the number of days elapsed from 6 A.M. 18th February, in the year 1612 of the Julian period, to the instant of the vernal equinox in A.D. 538, was 1329176 days, which being divided by the number of *Hindu* years, viz. 3639 then completed, we obtain $365^{d}\ 6^{h}\ 12^{m}\ 21^{s}\ 57^{th}$ or $365^{d}\ 15^{da}\ 30'\ 54''\ 54'''$ for the length of the *Hindu* year; which, however, must undergo a correction, in order to adjust it to the new system, under the following conditions:—1st, The *Kalpa*, or system, is to commence with *Sunday at sunrise, as the first day of the week.*—2d, *The number of days in the Kalpa, or in* 4,320,000,000 *years, must be complete without a fraction.*—3rd, *The number of days from the creation, or beginning of the Kalpa, to the* 18th *February, in the year* 1612 *of the Julian period, or in* 1972944000 *years, must be complete without a fraction.*—4th, *The days so elapsed, must, when divided by seven, leave a remainder of five, to indicate that the* 18th *February* 1612 *J. P. fell on Friday.* To comply with all these conditions, the length of the Hindu year, when corrected, comes out $365^{d}\ 6^{h}\ 12^{m}\ 9^{s}$, or $365^{d}\ 15^{da}\ 30'\ 22''\ 30'''$, differing a few seconds from the former. Commencing, therefore, this year on the 18th February 1612 J.P. at 6 A.M. the termination of the 3639th year falls on the 20th of March A.D. 538, at 53′ 51″ past 4 P.M. at which instant the following were the mean positions of the planets:—

Sun, European sphere,	11s	29°	25′	33″	Hindu sphere,	0s	0°	0′	0″
Moon, ...	1	17	46	32	....	1	18	20	59
Mercury, ...	5	5	44	45	....	5	6	19	12
Venus, ...	4	17	37	54	....	4	18	12	1
Mars, ...	9	1	42	31	....	9	2	16	58
Jupiter, ...	9	21	19	26	....	9	21	53	53
Saturn, ...	5	15	12	35	....	5	15	47	02

On the 18th February 1612 J.P. at 6 A.M. they were assumed to be in 0s, or the beginning of the Hindu sphere; therefore, to get their mean annual motions from this assumption, we must get their revolutions for the time elapsed, (3639 years complete,) and add them to the above positions, which will then give us the following:

Sun, ...	3639rev.	0s	00°	00′	0″
Moon, ...	48649	1	18	20	59
Mercury, ...	15109	5	6	19	12
Venus, ...	5915	4	18	12	21
Mars, ...	1934	9	2	16	58
Jupiter, ...	306	9	21	53	53
Saturn, ...	123	5	15	47	02

These quantities, being now divided by the time, 3639 years, we shall get the mean annual motions of each, as follow:

Sun, ...	1rev.	0s	0°	0′	0″
Moon, ...	13	4	12	46	30
Mercury, ...	4	1	24	45	1
Venus, ...	1	7	15	11	45
Mars, ...	0	6	11	24	5
Jupiter, ...	0	1	0	21	8
Saturn, ...	0	0	12	12	49

The mean annual motions thus deduced, as from two actual observations, would of course give the positions of the planets on the 20th March A.D. 538 at 53′ 51″ past 4 P.M. reckoning such motions as commencing at the epoch of mean conjunction. But, conformably to the nature of the system to be constructed, it is requisite that the planetary motions should commence with the *Kalpa*, or modern creation; therefore the motions just found will require a correction to adjust them to the system of years already mentioned. For this purpose, it is necessary, in the first place, that the number of revolutions of each planet in the period of 4320000000 years should be complete, and entire without a fraction; and, secondly, that the number to be assigned shall, when reckoned as commencing from the creation, or beginning of the *Kalpa*, give the mean place of the planet in A.D. 538, when the system was framed, as near the truth as the nature of integral numbers will admit. The method of doing this I will now show;—a single example will be sufficient for this purpose.

Let it be required to find from the mean motions above determined, the number of revolutions of Venus in the period of 4320000000 years, so that the same being reckoned from the creation, (1972947639 years before A.D. 538,) it shall give the mean place of the planet, sufficiently correct to answer all *Hindu* purposes.

First step.—The mean motion of Venus for 3639 years is found above to be $5915^{rev.}\ 4^{s}\ 18^{\circ}\ 12'\ 21''$. Therefore, as 3639 years give this quantity, so 4320000000 years will give 7022384850 revolutions nearly: then say,

As 4320000000 years to 7022384850 revolutions, so the years elapsed from the creation in A.D. 538 = 1972947639, to ($3207129076^{rev.}$) $4^{s}\ 16^{\circ}\ 24'\ 20''$; comparing this with $4^{s}\ 18^{\circ}\ 12'\ 21''$, the mean longitude of Venus in A.D. 538, it will be found too little by $1^{\circ}\ 48'\ 1''$; to make up this deficiency, we must find what difference one revolution will make, thus:—As $4320000000^{yrs.} : 1^{rev.} :: 1972947639^{yrs.} : 5^{s}\ 14^{\circ}\ 24'\ 44''\ 17'''\ 30^{iv.}\ 7^{v.}\ 12^{vi.}$ Having thus found the difference that one revolution would make, we must find by trials what number of revolutions will make up for the deficiency, 1° 48′ 1″;—this will be found to be 4642; for, if we multiply 5^s 14° 24′ 44″, &c. by this number, we shall get, rejecting the revolutions as of no use, 1° 54′ 42″, which exceeds 1° 48′ 1″ by only 6′ 41″: therefore, adding 4642 to 7022384850, we have the corrected number, equal to 7022389492, which is the very same that is given by the inventor of the system, in the following table; and in this manner all the rest were formed and adjusted to the number of years above given.

Table of the revolutions of the planets, apsides, and nodes in a *Kalpa*, or 4320000000 years, formed in the manner above explained by the author of the system in A.D. 538.

Planets.		Apsides.	Nodes Retrograde.
Sun's revolutions,	4320000000	 480	
Moon, ...	57753300000	 488105858	 232311168
Mercury, ...	17936998984	 332	 511
Venus, ...	7022389492	 653	 893
Mars, ...	2296828522	 292	 267
Jupiter, ...	364226455	 855	 63
Saturn, ...	146567298	 41	 584
Equinoxes, ...	199669	.. ..	.. ..
Days, ...	1577916450000	.. ..	.. ..

The following numbers, which are of use in computation, are derived from those in the above table.

Solar or Saura months, in 4320000000 years, ...		= 51840000000
Lunations	in ditto, = 57753300000 − 4320000000	= 53433300000
Intercalary lunations,	= 53433300000 − 51840000000	= 1593300000
Tithis, or lunar days,	= 53433300000 × 30	= 1602999000000
Intercalary Tithis,	= 1,602,999,000,000 − 1,577,916,450,000	= 25,082,550,000
Sideral days, ...	= 1577916450000 + 4320000000	= 1582236450000

Having thus given a complete view of the manner in which the first and most ancient of the modern astronomical systems of the *Hindus* was framed, it must, I believe, be sufficiently obvious to any person acquainted with computation, that the positions of the planets given by such system, must necessarily be nearer the truth at the time it was framed, than at any other distant period, either before or after. For, though entire numbers cannot be made to give exactly the positions of the planets according to observation, as may be seen by the example above given respecting the number of revolutions of Venus, which gives the position of that planet about 6′ 41″ too great, yet still, the errors upon the whole, will be less then than at any other distant time: and this self-evident principle will be found to exist, not only in *Hindu* astronomical works, but also in all astronomical tables whatever; for every astronomer, whether his system or tables be real or artificial, must necessarily endeavour to give the positions of the planets as correct as he can, at least in his own time; for otherwise they would be of no use.

Let us, therefore, now, apply this principle in determining the antiquity of the above system, in the same manner as if we met with it by accident, and did not know when it was framed. For this purpose there are two methods. The first is, to determine the error in the positions of the planets at some fixed point of time, and then to divide the errors so found, by the errors or differences in the mean motions; the mean result will point out the time sufficiently near for our purpose. The second is, to determine the errors in the positions of the planets at different periods, till one is found at which the sum of the errors is the least possible, and after which the errors again begin to increase. The period when the errors are least, is the time the system was framed, or very near it.

To determine the antiquity of the system by the first method, let the errors in the positions of the planets at the beginning of the *Kali Yuga*, (18th February 1612 J.P. at 6 A.M. 75° 50′ East of Greenwich,) be determined.

The positions of the planets by the system, at the beginning of the *Kali Yuga*, will be had by multiplying the four last figures of the number of revolutions of each, as given in the Table, by 4567, reserving in the product the four right-hand figures, which will express the position of the planet in decimal parts of a revolution, thus:—

The Sun, ...	0000 × 4567	=	0000					
Moon, ...	0000 × 4567	=	0000					
Mercury, ...	8984 × 4567	=	9928	=	11ˢ	27°	24′	28″,8
Venus, ...	9492 × 4567	=	9964	=	11	28	42	14 ,4
Mars, ...	8522 × 4567	=	9974	=	11	29	3	50 ,4
Jupiter, ...	6455 × 4567	=	9985	=	11	29	27	36 ,0
Saturn, ...	7298 × 4567	=	9966	=	11	28	46	33 ,6
Moon's apogee,	5858 × 4567	=	3486	=	4	5	29	45 ,6
—— Node, ...	1168 × 4567	=	4256	=	5	3	12	57 ,6

The rule above given, serving only to exhibit the positions of the planets, &c. at the beginning of the *Kali Yuga*, by the system, it may be proper to give here the general rule, which answers for all times. It is this: As 4320000000 years is to the number of revolutions of the planet, &c. given in the Table; so the time elapsed from the beginning of the *Kalpa*, or modern creation, to the planet's, &c. mean place in the *Hindu* sphere at the end of that time.* By this rule the above positions may be verified, the number of years elapsed at the beginning of the *Kali Yuga* being 1972944000. By the same rule the mean annual motions are also obtained, by making the statement for one year.

---

* If the time be days, you may make the days in the Kalpa your first number the revolutions the second, and the given number of days the third.

Having the positions of the planets at the beginning of the *Kali Yuga*, we compare them with their positions for the same instant by La Lande's Tables, in order to find the errors or differences at that time, thus:—

Mean places of the planets at the beginning of the *Kali Yuga*.

	By the system as above.				By La Lande's Tables.				Errors.			
Sun, ...	0s	0°	0′	0″	0s	0°	0′	0″		0°	0′	0″
Moon, ...	0	0	0	0	0	3	8	47,7	–	3	8	47,7
Mercury, ...	11	27	24	28,8	10	27	21	0,9	+	30	3	27,9
Venus, ...	11	28	42	14,4	1	2	52	50,9	–	34	10	36,5
Mars, ...	11	29	3	50,4	11	17	47	22,9	+	11	16	27,5
Jupiter, ...	11	29	27	36,0	0	16	49	20,7	–	17	21	44,7
Saturn, ...	11	28	46	33,6	11	8	46	40,0	+	19	59	53,6
Moon's apogee, .	4	5	29	45,6	3	29	58	19,2	+	5	31	26,4
—— Node supt.	5	3	12	57,6	5	6	38	3,8	–	3	25	6,2

Here, at the beginning of the *Kali Yuga*, we have the error in the place of Venus, equal to 34° 10′ 36″ 5, whereas in A. D. 538, it amounted to only 6′ 41″, which alone would be a convincing proof of the time when the system was framed; for all the errors must incontrovertibly diminish, as we approach the time at which the observations were made, on which the system is founded.

But to proceed: we must now compare the mean annual motions by the system, with La Lande's Tables for the same space of time, 365d. 6h. 12m. 9s., reduced to the Hindu sphere, in order to find the differences, or errors.

	By the System.				By La Lande's Tables.				Errors.	
Sun, ...	0s	0°	0′	0″	0s	0°	0′	0″	0″	
Moon, ...	4	12	46	30	4	12	46	26,6140	3,3860	+
Mercury, ...	1	24	44	59,6952	1	24	45	33,3660	33,6708	–
Venus, ...	7	15	11	56,8476	7	15	11	22,9260	33,9208	+
Mars, ...	6	11	24	8,5566	6	11	24	19,6790	11,1224	–
Jupiter, ...	1	0	21	7,9365	1	0	20	51,5178	16,4187	+
Saturn, ...	0	12	12	50,1894	0	12	13	10,4427	20,2523	–
Moon's apogee,	1	10	40	31,7574	1	10	40	36,5950	4,8376	–
——— Node,	0	19	21	33,3504	0	19	21	29,8975	3,4529	+

Having now obtained the errors or differences in the mean annual motions, let the errors in position at the beginning of the *Kali Yuga* be divided by these, and we shall have the time, according to each, when there was no error, thus:—

Moon, ...		3°	8′	48″	...... divided by	3,386	gives	3591
Sec. equat. +			13	54				
Mercury, ...	30	3	27	9	...... ditto by	33,66	—	3213
Venus, ...	34	10	36	5	....	33,92	—	3627
Mars, ...	11	16	27	5	....	11,12	—	3649
Jupiter, ...	17	21	44	7	....	16,42	—	3806
Saturn, ...	19	59	53	6	....	20,25	—	3555
Moon's apogee,	5	31	26	4	....	4,84	—	4109
—— Node supt.	3	25	6	2	....	3,45	—	3567
Their sum is					.. ..			29117

Which, divided by 8, gives for a mean result 3639 years from the beginning of the *Kali Yuga*, or A.D. 538, the real epoch of the system.

The following Table will now explain the second method, by which the errors in the system, at different periods, are shown gradually diminishing, from the beginning of the *Kali Yuga*, down to the epoch at which it was framed, A.D. 538.

Planets, &c.	Kali Yuga, errors at.	Kali Yuga 600.	Kali Yuga 1200.	Kali Yuga 2000.	Kali Yuga 3000.	Kali Yuga 3639, or A.D. 538.
Moon,*	8°55′33″ —	7° 8′14″ —	5°27′37″ —	3°26′11″ —	1°15′28″ —	0° 7′ 20″ +
Mercury,	30 3 28 +	24 26 46 +	18 50 3 +	11 21 6 +	1 59 56 +	3 58 40 —
Venus,	34 10 37 —	28 31 24 —	22 52 11 —	15 19 53 —	5 54 32 —	0 06 41 +
Mars,	11 16 28 +	9 25 15 +	7 34 1 +	5 5 43 +	2 0 21 +	0 1 54 +
Jupiter,	17 21 45 —	14 37 34 —	11 53 23 —	8 14 28 —	3 40 49 —	0 45 58 —
Saturn,	19 59 54 +	16 29 23 +	13 6 51 +	8 36 50 +	2 59 18 +	0 36 24 —
Moon's apogee	5 31 26 +	4 43 4 +	3 54 41 +	2 50 11	1 29 34 +	0 38 2 +
Moon's Node supt.	3 25 6 —	2 50 34 —	2 16 2 —	1 30 0 —	0 32 27 —	0 4 19 +

By this Table it may be seen, at one view, the vast difference there is between the errors at the be-

* Including the secular equation.

ginning of the *Kali Yuga*, and those in A. D. 538. At the former period, the error in the moon's place was 8° 55′ 33″, at the latter only about 7′ 20″, the secular equation being added throughout. The error in the place of Venus at the beginning of the *Kali Yuga* was upwards of 34°, in A. D. 538 only 6′ 41″ The error in Mars, at the former period, was upwards of 11°, at the latter only 1 54″; and so in all the rest, in not one of which does the error in A. D. 538 amount to a degree, except Mercury. The error in the place of Mercury in A. D. 538, I suspect, has not arisen from incorrect observation, but rather from some inadvertent error having crept into the number of revolutions by miscopying, at some period or other. The number of revolutions that seem requisite to correct the error is 300, which, being added to 17936998984, makes 17936999284. This number will give the place of Mercury in A. D. 538, agreeing with our modern European Tables within 17′ 13″

Having now sufficiently explained the structure of the astronomical part of the system, and the mode of determining the mean places of the planets, &c. from the number of revolutions of each in the *Kalpa*, it will be proper, in the next place, to say something of the Lunar Asterisms.

The Lunar Asterisms, from what has been said in the first part of this essay respecting ancient observations, must have existed ready formed, and the latitudes and longitudes of some of the principal fixed stars in each, determined many centuries anterior to A. D. 538: moreover, the Hindu sphere, at different times, appears to have commenced with different Asterisms, depending on the coincidence of the commencement of the year with that of the Lunar Asterism. Thus, in the year 1181 B. C. when the Hindu months were first framed and named, the year began with the month *Māgha*, at the winter solstice, and with the Lunar Asterism *Dhanisht'hā*, sometimes called *Sravisht'hā*, which was therefore made the first of the series at that time. But the ancient Hindu years, months, and seasons, being tropical, continually fell back in respect of the fixed stars; in consequence of which, at the end of every period of 247 tropical years and one month, the commencement of the year was changed to the next succeeding month, in regular succession, in the manner already described in the first part, until at length the month Vaisākha, in A. D. 538, coincided with the beginning of the Lunar Asterism *Aswinī;* which was therefore, made the first of the year, and *Aswinī* the first Lunar Asterism in the series of mansions: (see Plate VI.)* and this is the epoch referred to by the positions, &c. of the stars given in all the Hindu books written since that period.

It is proper to observe, that the positions of the stars usually given in Hindu books, are not, strictly speaking, either their latitudes and longitudes, nor declinations and right ascensions. The distance of the point in the ecliptic, cut by the circle of declination of the star, measured from the beginning of *Aswinī*, is given in place of the longitude; and the distance between the same point of the ecliptic and the star, measured on the circle of declination, is given in place of the latitude: from these the true latitude and longitude is obtained by computation when necessary. All this will be easily understood by means of the following Diagram: —

* In Plate VI. there are two numbers to each fixed Lunar Asterism, to point out the name: the first, or right hand number, refers to the modern order, beginning with *Aswinī;* the second to the ancient, commencing with *Sravisht'hā*. The moveable Lunar Mansions, after the introduction of the signs of the tropical sphere into India, I believe were discontinued, as of no further use; but they are marked in the plate, to point out the quantity of the precession, reckoning from the year 1192 B.C. when they coincided with the fixed or astral ones of the same name.

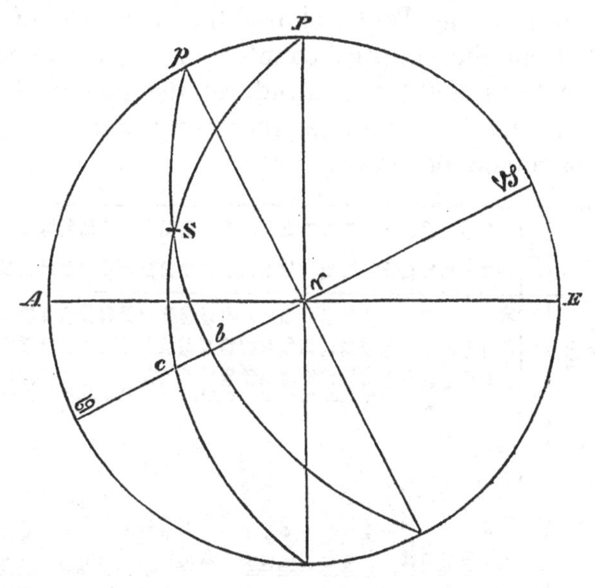

Let P, be the pole of the equator.
p, the pole of the ecliptic.
$A\,E$, the equator.
♋ ♈ ♑, the ecliptic.
S, the place of a star.
♈, the beginning of *Aswinī*, and the vernal equinoctial point in A.D. 538, then assumed as coinciding.

Then $P\,c$, is the circle of declination cutting the star at S, and the ecliptic in the point c. Now the distance ♈ c, on the ecliptic, is the tabular longitude from the beginning of *Aswinī*, and the distance c S, is the tabular latitude: on the other hand, ♈ b is the true longitude, because the circle of latitude $p\,b$, cuts the ecliptic at b, and the true latitude is b S.

The Hindu astronomers, in calculating the true latitudes and longitudes from the tabular ones, determine the difference of longitude $c\,b$, and add it to, or subtract it from the tabular longitude, according as the circumstance of the case may require.

The following Table, shows the distances of the stars from the ecliptic, counted on their circles of declinations, and the longitude of the points of the ecliptic cut by the same, reckoning from Aries, or the beginning of *Aswinī*.

Names of the Lunar Asterisms.	Distance S c	Dist. c ♈	Longitude of the star in the Mansion.	By Computation. Latitude or S b	diff. long. b c.	true long. ♈ b	Stars supposed to be intended.	Longitude in A. D. 1690.	Preces-sion.
1	2	3	4	5	6	7	8	9	10
1 *Aswinī*,	10° 0 N.	8° 0′	8° 0	9° 8′ N.	4° 5′ +	12° 5′	γ or β Arietis,	29° 14′ 29″	17° 9′
2 *Bharanī*,	12 0	20 0	6 40	11 11	4 41 +	24 41	35 Arietis,	42 35 47	17 54
3 *Kṛitikā*,	4 31	37 28	10 48	4 15	1 30 +	38 58	Alcyone,	56 40 8	16 42
4 *Rohinī*,	4 33 S.	49 28	9 28	4 22 S.	1 17 —	48 11	87 Tauri,	65 27 0	17 16
5 *Mṛigasiras*,	10 0	63 0	9 40	9 48	2 0 —	61 0	113, 116, 117 Tauri?	77 50 11	16 50
6 *Ārdrā*,	11 0	67 0	0 20	10 50	1 55 —	65 5	133 Tauri?	82 38 44	17 33
7 *Punarvasu*,	6 0 N.	93 0	13 0	6 0 N.	0 8 —	92 52	β Geminorum,	108 56 9	16 4
8 *Pusyā*,	0 0	106 0	12 40	0 0	0 0 —	106 0	δ Cancri,	124 23 40	18 24
9 *Asleshā*,	7 0 S.	108 0	1 20	6 56 S.	0 58 +	108 58	49, 50 Cancri,	126 32 0	17 34
10 *Maghā*,	0 0 N.	129 0	9 0	0 0 N.	0 0 —	129 0	Cor Leonis,	145 31 20	16 31
11 *P. Phalgunī*,	12 0	147 0	10 40	11 14	4 15 —	142 45	70, 71 Leonis,	158 46 0	16 1
12 *U. Phalgunī*,	13 0	155 0	8 20	12 2	4 56 —	150 4	β Leonis,	167 19 14	17 15
13 *Hastā*,	11 0 S.	170 0	10 0	10 4 S.	4 28 +	174 28	7, 8 Corvi,	189 20 0	14 52
14 *Chitrā*,	2 0	183 0	9 40	1 50	0 49 +	183 49	Spica Virginis,	199 31 21	15 42
15 *Swāti*,	37 0 N.	199 0	12 20	33 41 N.	16 19 —	182 41	Arcturus,	199 53 52	17 13
16 *Visākhā*,	1 23 S.	212 5	12 5	1 18 S.	0 29 +	212 34	24 Libræ,	226 41 43	14 8
17 *Anurādhā*,	1 44 N.	224 5	10 45	1 39 N.	0 28 —	223 33	β Scorpii,	238 52 56	15 20
18 *Jyesht'hā*,	3 30 S.	229 5	2 25	3 22 S.	0 59 +	230 4	Antares,	245 26 4	15 22
19 *Mulā*,	8 30	241 0	1 0	8 19	1 48 +	242 48	34, 35 Scorpii,	259 58 13	17 10
20 *P. Āshād'hā*,	5 20	254 0	0 40	5 18	0 0	254 0	δ Sagittarii,	270 14 12	16 14
21 *U. Āshād'hā*,	5 0	260 a	6 40	4 59	0 32 +	259 37	φ Sagittarii,	275 50 42	16 14
* *Abhijit*,	62 0 N.	265 b	11 40	61 56 N.	4 14 —	260 46	α Lyræ,	280 57 18	20 17
22 *Sravanā*,	30 0	278 c	11 20	29 56	2 3 +	280 3	α Aquillæ,	297 23 24	17 20
23 *Dhanisht'hā*,	36 0	290 d	10 0	35 32	6 15 +	296 15	α Delphini,	313 3 24	16 48
24 *Satabhishā*,	0 18 S.	320 0	13 20	0 17 S.	0 6 —	319 54	λ Aquarii,	337 14 41	17 20
25 *P. Bhādrapadā*,	24 0 N.	326 0	6 0	22 26 N.	8 46 +	334 46	α Pegasi,	349 9 13	14 23
26 *U. Bhādrapadā*,	26 0	337 0	3 40	23 56	10 29 +	347 29	γ Pegasi?	4 49 50	17 21
27 *Revatī*,	0 0	360 0	13 20	0 0	0 0	360 0	ζ Piscium.	15 32 13	15 32

The first column of the above table shows the names of the Lunar Asterisms in their order, reckoning from *Aswinī* as the first of the series. The second column contains the distances of the respective stars from the ecliptic, measured on their circles of declinations, explained in the diagram by the distance S *c*. The third column shows the longitude of the point in the ecliptic, intersected by the circle of declination (at *c* in the diagram), measured on the ecliptic, and represented by ♈ *c*.

The fourth column shows the longitude of each star in its own mansion.

The fifth, shows the true latitude, marked in the diagram S *b*, and the sixth, the difference of longitude *b c*, both computed from the positions given in the 2nd and 3rd columns.

The seventh, shews the true longitude represented by ♈ *b*.

The eighth, the stars supposed to be intended by the positions so given.

The ninth, the longitudes of the same stars by the Britannic Catalogue in A. D. 1690, and thence we obtain the precession contained in the tenth column for each particular star, from the time of supposed observation down to that year.

The change took place in the Hindu astronomy in A.D. 538: from thence to A.D. 1690 are 1152 years, the precession for which is about 16° 2′ 4″. This, compared with the precessions in the 10th column, shows that some of the observations, if made in A.D. 538, must have been inaccurate, or else that they had been made at different times, and introduced into the table without correction. This seems to be the case with the star α Lyræ *(Abhijit)*, against which stands the precession 20° 17′, which exceed the real precession by about 4° 1-4th; and therefore supposed to be introduced into the table without making the necessary correction, to reduce its position to what it should have been at the epoch. It is, however, to be observed, that the star *Abhijit,* or rather the mansion so called, does not belong to the division of 27, but to that of 28: and the same may be said of the asterisms marked 21, 22, and 23, the positions of the stars given in them, from this cause, falling into mansions bearing other names, in the division of 27, which will be more fully explained in the next section.

The positions of the stars thus intended for the epoch of A. D. 538, are nearly the same in all Hindu books, whether written then, or at any time since; in order that the precession may be invariably reckoned from the commencement of *Aswini,* which is the beginning of the modern Hindu sphere. They, therefore, do not point out the age of the book in which they are given, which may be very modern, but only refer to the epoch of the change in the Hindu astronomy: so that whatever antiquity may be feigned to a Hindu book containing the positions of the stars thus given, or the same order of the mansions, we may be certain that it has been written since the introduction of the modern astronomy. So far, therefore, the table of mansions may be of use in limiting the utmost age of a book, when we have no other means to fix or determine its real date.

Having thus explained the Lunar Asterisms, it will be now proper to add a few remarks on the signs, both in the *Hindu* and tropical spheres. I have already noticed, that the *Hindus,* for some centuries anterior to A. D. 538, adopted the tropical sphere; or that in which the sign named Aries, always begins at the vernal equinoctial point, and which probably they received from the west. In A.D. 538, they changed their method, and introduced the sideral sphere now in use, which they divided into twelve signs of two Lunar Asterisms and a quarter each, the first of which always begins with *Aswinī*, and therefore fixed, in respect of the Lunar Asterisms; but the tropical signs, which they found necessary still to retain for a variety of purposes, continually falling back in respect of the others, by reason of the precession, it became, therefore, necessary to distinguish them from each other, in order to guard against confusion or uncertainty: this was effected by retaining the names Aries, Taurus, Gemini, Cancer, Leo, Virgo, &c. exclusively for the signs of the tropical sphere, and the new sideral signs, to be only numerically expressed or designated. Thus, suppose the sun's place was $9^s\ 6^\circ\ 4'\ 30''$, it would be immediately understood that this was in the sideral sphere, reckoning from the beginning of *Aswinī*, and not from Aries, or the vernal equinoctial point. For, if the latter was intended, it would be expressed thus: *Capricorn*, $6^\circ\ 4'\ 30''$, or else by the words *with the precession added*, when the name of the sign was not used. This method was also adopted in Europe about 150 years ago, when some astronomers had introduced the sideral sphere, making γ Arietis the commencement, and from which star the precession was reckoned, as by the *Hindus*, from the beginning of *Aswinī*.

The *Hindu* astronomers employ the tropical sphere to this day for many purposes. By the sun's longitude in the tropical sphere, or from the beginning of Aries, is determined his declination and right ascension, length of the day and night, times of sun's rising and setting, &c. together with the times of the heliacal and cosmical risings and settings of the stars, nativities, and a great many other circumstances known to astronomers. I have been more particular on this head, perhaps, than may appear necessary to the real astronomer; but my reason and excuse for it is, that in no part of the *Hindu* astronomy has arisen so much error and confusion as on this very point.

The astronomical system above explained, is by some attributed to Brahma, by some to Brahma Achārya, and by others to Brahma Gupta, the whole of which names, I apprehend, belong to one and the same individual, and that individual to be Brahma Gupta. For the system is given in the *Siddhanta Siromani*, said to be by Bhāskara Achārya, and acknowledged to be from Brahma Gupta; and the same conclusion is supported by the authority of other writers, notwithstanding the opinion of the commentator on the *Surya Siddhanta*, that Brahma Gupta borrowed his system from the *Vishnu Dharmottara Purāna*, an opinion which can be of no weight whatever; because, it is the wish of every *Hindu*, to make the world believe in the great antiquity of their Purānas: though in fact none of them are ancient, and some of them not a hundred years old. Brahma Gupta's system may be contained in the *Vishnu Dharmottara Purāna*, but it does not follow from thence that he borrowed it from that work; on the contrary, it is more reasonable to suppose that the author of the *Vishnu Dharmottara Purāna* borrowed it from him.

But be this as it may, the question who was the author of it, is not of the slightest importance; nor do I care or concern myself about who was the author: my object alone was to determine, from astronomical data, the antiquity of the system, which, I believe, I have sufficiently and satisfactorily effected.

The object of the author of the system, whoever he may have been, was evidently to substitute in the room of the former periods (the four ages and nine patriarchal periods, or Manwantaras), the immense periods of his own system, and thereby give the appearance of the most profound antiquity to the *Hindu* people, their history, their arts, and their sciences, far beyond any other nation or people on the face of the earth, as may be seen by the following passage in the Commentary on the *Surya Siddhanta*, wherein Ganesa is made to say: "The planets were right in the computed places in the time of Brahma Acharya, Vasishtha, Casyapa, and others, by the rules they gave; but in length of time they differed, after which, at the close of Satya age (2,163,101 years before Christ), the sun revealed to Meya a computation of their true places. The rules then received answered during the *Tretā* (1,296,000 years), and *Dwāpar* (864,000 years), as also did other rules formed by the *Munis* during those periods. In the beginning of the *Kali Yuga* (3101 B.C.) Parāsara's book answered; but Āryabhatta, many years after, having examined the heavens, found some deviation, and introduced a correction of bija. After him, when further deviations were observed, Durgā, Sinha Mihira, and others, made corrections. After them came the son of Jistnu, and Brahma Gupta, and made corrections. After them, Kesava settled the places of the planets; and sixty years after Kesava, his son Ganesa made corrections."—*As. Res.* vol. ii. p. 243.

The object of this most absurd passage is, first, to give an appearance of immense antiquity to their astronomy, their history, and consequently themselves as a people; secondly, to throw back into antiquity Āryabhatta, Durgā Sinha, Varāha Mihira, and others, by placing Brahma Gupta as posterior to all these; which, however, is shown to be false in the very beginning of the book, where the author enumerates the astronomers whose works he consulted, in the order of their antiquity, thus: Brahma Gupta, Āryabhatta, Varāha, Lalla, &c. The real times of Āryabhatta and Varāha Mihira will be shown from their own works, in the third and fourth sections: so that the above passage must be an imposition, if Brahma Gupta was the author of the system above given, or lived at the time when it was framed.

In fact, there is no imposition too gross or absurd that a *Hindu* will not employ to gain his ends, if he can effect it by that means. We see that by means of this system of Brahma (invented in A.D. 538), and of various passages like the above, inserted in books with a view to support it, the real *Hindu* history and chronology have been completely destroyed; so that Yudhisht'hira, Parāsara, Garga, and others, who lived from about 540 to 575 B.C. were thrown back into antiquity about 2600 years more:* Rama, who was born in the year B.C. 961, was thrown back upwards of 867,000 years before the Christian era, and Ikswaku, the first king, who began his reign in the year 1528 B.C. was thrown back upwards of 2,163,000 years B.C.; for such was the change made by this system in the chronology and history of the *Hindus*. But to carry all this into effect, many things were

* This is the cause why the reigns of the kings from 540 B.C. to 299 B.C. are not distinguished from the rest that followed. See pp. 77, 78.

necessary. In the first place, it was requisite that all their ancient books on astronomy, history, &c. that could in the smallest degree affect or contradict the new order of things, should be either destroyed new modelled, or the obnoxious passages expunged; and, secondly, that others should be written or composed having the appearance of antiquity, by being fathered on ancient writers to support, as it were by their evidence, the existence in ancient times, and through all ages, of the new system of years thus introduced. Thus, it is put into the mouth of Menu to say: "When ten thousand and ten years of the *Satya Yuga* were past (i. e. 3881091 B.C.) on the night of the full moon in the month of *Bhādra*, I Munnoo, at the command of Brahma, finished this *shaster*, that speaks of men's duty, of justice, and of religion, ever instructive." * By such means the system was introduced, though I believe not without a struggle; for there is still a tradition that the *Maharastras* or *Maharattas*, destroyed all the ancient works,—that people hid their books in wells, tanks, and other places, but to no purpose, for hardly any escaped; and those that did then escape, were afterwards picked up by degrees, so that not one was allowed to be in circulation. This will account, not only for the books that now exist being either entirely modern, or else new modelled to correspond with the new order of things, but also for the paucity of ancient facts and observations that have reached our time. Indeed the few scattered and insulated fragments that have reached our time,

* Mr. Halhed, in his Gentoo laws, gives this passage, and from what he says, seems to believe the truth of it; but his argument will not hold good, because it does not follow that Menu ever wrote a line of it: in fact he could not, Menu being only a fictitious personage introduced for the purpose.

would not have been allowed to pass, had they been supposed to be of any consequence, or could convey any idea or knowledge of former times. And we may rely on it, that the moment they become known, the books in which they are contained will either be destroyed, or the facts themselves expunged; for the *Brahmins* of this day, are fully as eager in support of this monstrous system as those that first invented it, and wàtch every opportunity of destroying such facts against it as may appear to have escaped the vigilance of former *Brahmins*. But to wait for the gradual development of facts, would be a great loss of time: they therefore artfully endeavour, as if by accident, to encourage a controversy on the subject, with the sole view of knowing from the opponent, the points on which he rests his arguments, and the books from whence he draws them, in order that such books may be destroyed entirely, or the facts expunged by degrees, as the nature of the case will admit. It is but too well known, that many books that were in circulation not more than fifty years ago, have now altogether disappeared, probably from this cause alone.

To some it would doubtless appear, as a thing impossible, that a set of *Brahmins* in *Ujein*, could impose such a system on the rest of India. Those, however, who are acquainted with the *Brahminical* character, know too well that every thing was in their power: they were in possession of all the learning in the country, and their influence was so great, that even the princes of the country were obliged to bow submission to their will. Therefore, when they assembled together in convocation, to consult on the general interest of the whole body, whatever resolutions they came to on that head, would be universally adopted by the brethren: and woe to the man that should dare oppose them; for their power and influence far exceeded those of the Popes in Europe, so that wherever they sent their secret orders, they would be sure to be obeyed.

The introduction of the modern system was doubtless intended as a blow on Christianity, which, at the time, was making some progress in *India;* for by making the Christians appear but as people of yesterday, in comparison to themselves, the natives would not only be less disposed to listen to them, but would look upon them with the same degree of contempt as the *Brahmins* did.

But the grandest blow of all, which was levelled by the *Brahmins* against Christianity, and the *ne plus ultra* of their schemes, was the invention of the *Avatars,* or descents of the Deity, in various shapes, and under various names, particularly that of Krisna; for as the Christians acknowledged that Christ was an incarnation of the Deity, and that God the Father had sent him down on earth to show his special favour to them, and redeem them from sin; so the *Brahmins,* in return, invented not one, but several incarnations and descents of the Deity amongst them at various times; thereby, to make it appear by such frequent descents, that they exceeded the Christians and all other nations by far, in point of favour with the Deity.

My attention was first drawn to this subject, by finding that a great many of the *Hindu* festivals marked in their calendar, had every appearance of being modern; for they agreed with the modern astronomy only, and not with the ancient.

I observed also several passages in the *Geeta,* having a reference to the new order of things. I was therefore induced to make particular enquiries respecting the time of Krishna, who, I was satisfied, was not near so ancient as pretended. In these enquiries I was told the usual story, that Krishna lived a great many ages ago; that he was contemporary with Yudhishthira; that Garga, the astronomer, was his priest; and that Garga was present at his birth, and determined the positions of the planets at that moment; which positions were still preserved in some books, to be found among the astronomers: besides which, there was mention made of his birth in the *Harivansa* and other *Purānas*. These I examined, but found they were insufficient to point out the time. I therefore directed my attention towards obtaining the *Janampatra* of Krishna, containing the positions of the planets at his birth, which at length I was fortunate enough to meet with, and which in the original Sanscrit runs in the words following: —

उच्चस्थाः शशिभौमचान्द्रिशनयो लग्नं वृषो लाभगो जीवः
सिंहतुलालिषु क्रमवशात् पूषोशनोराहवः। नैशीथः समयो
ऽष्टमी ब्रह्मर्क्षमत्रक्षणे श्रीकृष्णाभिधमम्बुजेक्षण मभूदाविः
परं ब्रह्मतत्॥

From which it appears that Krishna was born on the 23d day of the moon of *Srāvana,* in the Lunar Mansion *Rohinī,* at midnight; at which instant the moon, Mars, Mercury, and Saturn, were in their respective houses of exaltation; the moon in Taurus, Mars in Aries, Mercury in Virgo, and Saturn in Libra: that the sign Taurus was then rising: Jupiter in Pisces, the sun in Leo, Venus and the moon's ascending node in Libra.

The positions of the planets thus given us at the birth of Krishna, place the time of the fiction to the year A. D. 600, on the 7th of August, on which day, at noon, on the meridian of Paris, the following were their respective positions, as computed from European Tables:—

Sun in Leo,	4^s	16°	40′
Moon in Taurus,	1	18	32
Moon's node asc. in Libra,	6	11	17
Mercury in Virgo, geoc. long.	5	0	29
Venus in Libra, ... do.	6	1	24
Mars in Aries, ... do.	0	16	46
Jupiter in Taurus, ... do.	1	10	5
Saturn in Libra, ... do.	6	26	51

Subtracting the sun's longitude, 4^s 16° 40′, from the moon's, 1^s 18° 32′, we get 9^s 1° 52′, which being divided by 12, the difference in longitude between the sun and moon in a lunar day, we have 22 lunar days, 29 dandas, and 20′, and therefore only 20 dandas 40′ to the commencement of the next lunar day, or about 8^{hs} and 24 minutes, making the commencement of the 23d at 24^m past 8 in the evening. To this add difference of meridians, 4 hours, 54 minutes, makes at Ujein, 18^m past one in the morning, at which time the moon was a little past the middle of *Rohinī*.

Krishna, as a portion of Vishnu, means time, or the year; for Vishnu being a personification of time, any portion whatever of him must be considered as time also. Hence the figure of Krishna is almost always accompanied with that of one or more serpents, as emblematic of time; for all the deities whose representations or sculptures are accompanied by figures of serpents, are without doubt mere personifications of time, whether taken as the year or time indefinite*. Arjuna in the *Geeta*, page 93, in addressing Krishna, says: "I am anxious to learn thy source, and ignorant of what thy presence here portendeth." Krishna answers: "*I am Time, the destroyer of mankind, matured, come hither to seize at once on all these who stand before us.*"

The fabrication of the incarnation and birth of Krishna, was most undoubtedly meant to answer a particular purpose of the *Brahmins*, who probably were sorely vexed at the progress Christianity was making, and fearing, if not stopped in time, they would lose all their influence and emoluments. It is, therefore, not improbable but that they conceived, that by inventing the incarnation of a deity nearly similar in name to Christ, and making some parts of his history and precepts agree with those in the gospels used by the Eastern Christians, they would then be able to turn the tables on the Christians by representing to the common people, who might be disposed to turn Christians, that Christ and Krishna were but one and the same deity; and as a proof of it, that the Christians retained in their books some of the precepts of Krishna, but that they were wrong in the time they assigned to him; for that Krishna, or Christ, as the Christians called him, lived as far back as the time of Yudhishthira, and not at the time set forth by the Christians. Therefore, as Christ and Krishna were but one and the same deity, it would be ridiculous in them, being already of the true faith, to follow the imperfect doctrines of a set of outcasts, who had not only forgotten the religion of their forefathers, but the country from

---

* All the Hindus are Saturnalians, that is, worshippers of time, under various shapes and names, according to the different sects.

which they originally sprung. Moreover, that they were told by Krishna, in his precepts, that a man's own religion, though contrary to, is better than, the faith of another, let it be ever so well followed. "It is good to die in one's own faith; for another's faith beareth fear." *Geeta*, pp. 48, 49.

I have thus endeavoured to explain, what I conceive the motives of the *Brahmins* to have been, in their invention of the incarnations of Vishnu, particularly that of Krishna: nor have I any doubt but that the whole of the incarnations were invented at one and the same period; and as they were then destroying the old, and forging new books, to answer the purpose of the newly introduced system above explained, an opportunity offered of referring them to different portions of history, that the whole might have the appearance of reality. Krishna they artfully threw back to the time of Yudhishthira, because by that means they put the matter beyond the power of investigation, following exactly the examples of the Egyptians, Chaldeans, and Greek priests and poets, in throwing back the times of the war between the gods and giants, the Argonautic expedition, and the war of Troy, to periods of time out of the power of any one to contradict them: and this in fact is the case with almost all fictions, however plausible they may be.

What shall be now thought of the antiquity of *Hindu* books, in most of which, and particularly the *Mahābhārat*, they give the exploits of Krishna? Even some of the *Vedas* speak of him, which certainly is not saying much in favour of their antiquity. The age of the *Mahābhārat* is mentioned in Sir Stamford Raffles' History of Java, by which, at the utmost, it could not have been written earlier than about the year 786 of the Christian era; but from the words forming the date, the probability is, that it was as late as the year A.D. 1157. The war of the *Mahābhārat,* most likely, is nothing more than a mere fiction of the poet.

It is somewhat remarkable that none of the writings of the Christians who resided in India anterior to, or about A.D. 538, have come down to our time. If any exist at this day, they would most probably be invaluable, in throwing light on that part of the Hindu history, &c. which is now lost, in consequence of being either destroyed or concealed by the moderns, to make room for their new system. It is not impossible, however, but that some of them, with early Hindu manuscripts, may still exist, locked up in some immense public or private libarary in Europe, totally unknown and forgotten; where they will remain, until the great Hindu deity, TIME, puts an end to them, by finally mouldering them into dust.

SECTION II.

System of Varāha—*Framed in the ninth century—The object of it—Works in which it is given—Observation on Canopus referring to A. D.* 928—*Revolutions of the Planets, &c. in the System—Years elapsed to the beginning of the Kali Yuga—Formation of the System, with Remarks—Compared with the System of* Brahma—*Age of the System determined—Lunar Asterisms—The places of some Stars not agreeing with the names of the Mansions—The cause explained, and shown in a Table—Precession of the Equinoxes—The method employed artificial, by assuming the motion in an epicycle—Explained by a Diagram—The terms Libration or Oscillation inconsistent with the Author's meaning, which is further explained by the Commentator, &c. &c.*

We now come to notice the next astronomical system of the Hindus, in point of antiquity; that is to say, the system of Varāha. This system, from the astronomical data it gives, appears to have been formed in the ninth century of the Christian era. The object of the author, whoever he may have been, was, first, to support the notions introduced by the last system, in respect to the time of the creation, &c. &c.; secondly, to give new numbers that would give the positions of the planets correctly at the time, those given in the former system no longer answering, with sufficient accuracy, that purpose; and, thirdly, to render the calculation of the places of the planets, &c. much more simple and easy by smaller numbers, than could be done by the unwieldy numbers in the system of Brahma.

The system of Varāha* is given in the following

* This name Varāha, is supposed by some to have an allusion to the feigned incarnation of Vishnu under that name, while others suppose it to be from Varāha Mihira; but it is perfectly immaterial which. It is certain, however, that Varāha Mihira was not the author of the system, as will be seen in the fourth Section, where

works:—the *Vasisht'ha Siddhanta,* the author of which is pretended to have lived 1299101 years before the Christian era; the *Surya Siddhanta* pretended to have been written 3027101 years before the Christian era*; and the *Soma Siddhanta,* feigned to have been written by Gopi Raja at the close of the Dwāpar, or 3101 years B.C.

By these may be seen, the mode adopted for supporting the imposition introduced by the former system, that of people living, and cultivating the arts and sciences at immense periods of time back.

The three works just mentioned, it is probable, were written at different times, as occasion required, to support the imposition. The *Vasisht'ha Siddhanta,* I consider as the oldest, because the supposed author of that work is said to have observed Canopus, when that star was exactly in the beginning of Cancer.

This observation is mentioned by Dādā Bhāi, a commentator on the *Surya Siddhanta,* and is the very position given in the *Surya Siddhanta,* which we must therefore conclude to have been written subsequent to the observation of Vasisht'ha.

---

it will be shown that he was contemporary with Akber. This, however, does not prove that the system might not be called after him; neither does it prove any thing against his being the supposed author of the *Vasisht'ha* the *Surya,* and *Soma Siddhantas,* under feigned names, &c. the better to support the modern system of years introduced in A. D. 538; but whether he was or was not, is of no consequence whatever: we can ascertain pretty nearly the age of any system of astronomy, if genuine, though we cannot tell who framed it; neither can we tell who has been the author of any book, where the name given, as generally the case, is fictitious. What gave rise to the idea of Varāha being the author of the books just mentioned was, that the system they contained was called that of Varāha, a name naturally supposed to be given to the system, in consequence of his being the real or supposed author of these works. Systems may have no names conferred on them for many centuries after they are framed, as is the case with that given by Aryabhatta, Sec. III. which, I believe, to this day has no particular name assigned to it.

* The author pretends to have written it at the vernal equinox, beginning of the *Satya Yuga,* or 3027101 years before Christ.

The longitude of Canopus in A. D. 1750 was....	3ˢ	11°	30′	39″	6
The difference of longitude since the observation,	0	11	30	39	6

Which reduced to time, at 1° in 71½ years, we get 822 years, which being taken from 1750, leaves A.D 928, the time of the observation. From this fact, supposing we had no other, it would most undoubtedly appear that the *Surya Siddhanta* could not be written earlier than the 10th century. But besides the observation above given, we have also the time from the positions of the planets; which prove, that the system of Varāha, is even posterior to the observation on Canopus by the supposed Vasisht'ha. Before we proceed, however, to show the time from the positions of the planets, it is proper that we should first exhibit the system, and explain its structure.

The following Table exhibits the revolutions of the planets, &c. in 4320000000 years according to the system:—

Names.	Revolutions.	Apsides.	Nodes.
Sun, ...	.. 4320000000	Revol. 387	
Moon, ...	.. 57753336000	.. 488203000	Revol. 232238000
Mercury, ...	.. * 17937024000	.. 386	.. 488
Venus, ...	.. 7022376000	.. 535	.. 903
Mars, ...	.. 2296832000	.. 204	.. 214
Jupiter, ...	.. 364220000	.. 900	.. 174
Saturn, ...	.. 146568000	.. 39	.. 662
Days, ...	.. 1577917828000		
Revolutions of the equinoxes (in the epicycle,) 600000			

The number of years in this system, is exactly the same as in that of Brahma; but they do not commence at the same time. The system or *Kalpa*

* This is the correct number: in the system it is 17937060 for a *Mahā Yuga*, owing to an error in the number for the primary cycle, which should have been 4484256, and not 4484265. The digits are the same, but the two last are misplaced; the 6 should follow the 5.

of Varāha begins later by 17064000 years, a circumstance owing to the formation of the revolutions of the planets into small cycles, for the convenience of calculation. Therefore, in computing the number of years elapsed of this system, the time must be first found according to the system of Brahma, as already shown, and from that time 17064000 must be deducted, to give the years elapsed of the system, of Varāha. Thus, at the beginning of the *Kali Yuga*, there were elapsed of the system of Brahma, . . . 1972944000 years.
Deduct the above number, 17064000
Remain time elapsed of the system of Varāha, . 1955880000

Hence, it must be obvious, that the system of Brahma had existed and was in use long before that of Varāha; as the computation of the time elapsed must be made in the first instance by the former system, otherwise we should not know when the latter began.

The *Kalpa* of Varāha begins with Sunday, as the first day of the week, at the instant of midnight, on the meridian of *Ujein;* and the *Kali Yuga* begins with Friday at midnight. The year, therefore, begins earlier by six hours than in the system of Brahma, which would therefore cause it to be something longer; but the true length depends on other circumstances.

The revolutions of the planets given in the Table, all terminate with three cyphers: these being cut off, the remainder will be the revolutions in a *Mahā Yuga*, or 4320000 years. The numbers may be further reduced by dividing them by four, the quotient will be the revolutions in 1080000 years; which is the least common cycle, in which the planets return to a line of mean conjunction in the beginning, both of *Aswinī* and Aries.

The years elapsed of the system of Varāha, at the beginning of the *Kali Yuga*, are, as above, 1955880000. If this number be divided by the years in the least common cycle, 1080000, the quotient will be 1811, the number of cycles in that period. Now since the system begins with a Sunday, as the first day of the week, and the *Kali Yuga* begins with Friday, it is evident that the whole number of days to be assigned to the 1811 cycles, must, when divided by seven, leave a remainder of five: then the question is, how many odd days, over and above complete weeks, we must assign to each cycle, so as to answer this purpose. This is easily known by assuming one day in excess above the weeks: then 1811 cycles will have 1811 days, which, being divided by seven, leaves a remainder of five, which is the very number we want. Therefore, each cycle must contain a complete number of weeks, and one day over.

The time elapsed from the beginning of the *Kali Yuga* at midnight, to the instant of the vernal equinox in A. D. 538, which was supposed, or assumed to coincide with the beginning of the Lunar Asterism *Aswinī*, is 1329176 days, 6 hours, and 40 minutes in 3639 years: therefore, as 3639 years are to 1329176 ^{days} 6^{h} 40^{m}, so 1080000 years to 394479356 days, rejecting fractions. If we divide this number by seven, there will be a remainer of five; but from the conditions already stated, it must be one: therefore we add to the number three days more, and make it 394479359, from which the length of the year would be obtained, that would answer all the terms, because divisible by seven, leaving a remainder of one. But the number of days to be assigned to the cycle of 1080000 years, must also give the relative motions of the sun and moon correct, and the lunation of a true length; in consequence of which, a further correction must be made to the time by the addition of 14 complete weeks, or 98 days, to the above number, which will make it 394479457 days. This number being multiplied by four, makes 1577917828, the number of days in the *Mahā Yuga,* or 4320000 years; and the days in a *Mahā Yuga* being multiplied by 1000, the product will be 1577917828000, the days in a *Kalpa,* or 4320000000 years, the same as in the Table: and the days in either of the periods being divided by the corresponding number of years, we get the adjusted length of the Hindu year, according to the system of Varāha, $=365^{ds}$ 15da 31° 31′ 24‴. The length of the year might be computed in another manner; but as the result would be the same, by reason of the adjustment, it was thought unnecessary. Having found the length of the adjusted Hindu year, we are now enabled to proceed to show how the revolutions of the planets have been obtained:—one example will be perfectly sufficient.

I have shown above, that the observations of Vasishtha on the position of Canopus, refer us about to the year A. D. 928, or of the *Kali Yuga* 4029. Let us, therefore, suppose the system was framed about that period, and that the author had determined, by accurate observations, the positions of the planets in the Hindu sphere, at the end of the year 4029 of the *Kali Yuga;* which position formed the basis of the revolutions in the system. Thus, suppose we take Venus as an example:—

The mean heliocentric longitude of Venus at the end of the year 4029 of the *Kali Yuga,* A. D. 928,

March 23d. at 19^h 48′ 56″, meridian of Paris, by La Lande's Tables European sphere, was 4^s 9° 30′ 25″
Or Hindu sphere, ... 4^{sgs} 2 49 39
Now suppose this to have been the precise position determined by the author, it is required to determine the number of revolutions of the planet in the primary cycle of 1080000 years, that will give this position, reckoning from midnight at the beginning of the *Kali Yuga*, as an epoch of mean conjunction of all the planets.

The first step is to ascertain the number of revolutions made by the planet Venus, from the instant of midnight at the beginning of the *Kali Yuga*, to the end of the year 4029, which will be found to be 6549. Add these to the position of the planet at the end of the year 4029, $=4^s$ 2° 49′ 39″, the sum will be Venus's entire motion in 4029 Hindu years, $=$ $6549^{rev.}$ 4^s 2° 49′ 39″; then say, as 4029 years give $6549^{rev.}$ 4^s 2° 49′ 39″, so 1080000 years will give $1755594^{rev.}$ 0^s 22° 55′ 8″, which, rejecting the fraction, (being under six signs), it will be 1755594 revolutions in 1080000 years: this number being multiplied by four, we have 7022376, the revolutions in a *Mahā Yuga*, or 4320000 years; and the revolutions in a *Mahā Yuga* being multiplied by 1000, we have 7022376000, the revolutions in a *Kalpa*, the same as given in the Table.

Having thus shown how the number of revolutions of the planet in the primary cycle of 1080000 years is found, let it now be applied in determining the position of the planet. In doing this, there is no occasion to reckon from the beginning of the *Kalpa*, as in the system of Brahma; for in this system, we may commence our calculation from any point of time, at which the planets are assumed to have been in a line of mean conjunction: and as the last mean conjunction is assumed to have taken place at midnight, at the beginning of the *Kali Yuga*, we commence our calculation from thence, in preference to any other, as being the commencement of the cycle of 1080000 years.

Suppose we wanted to know the mean heliocentric longitude of Venus at the end of the year 4029 of the *Kali Yuga*, (A.D. 928,) then we say, as the number of years in the cycle, is to the number of revolutions in the same, so the years elapsed since the beginning of *Kali Yuga*, to the planet's heliocentric longitude, in revolutions, signs, &c. Thus, for the year 4029 of the *Kali Yuga*, we have $\frac{1755594 \times 4029}{1080000} = 6549^{rev.}\ 4^{s}\ 2^{\circ}\ 44'\ 31''\ 12'''$, which differs from the actual position of Venus by only $5'\ 7''\ 48'''$: this difference arises from the fraction $0^{s}\ 22^{\circ}\ 55'\ 8''$, being rejected in forming the number of revolutions, which must be always entire.

The motions of the nodes and apsides of the planets being slow, making a revolution in a great many years, are reckoned from the beginning of a system, and their revolutions determined in the same manner as those of the planets in the system of Brahma, already explained.

It must be obvious from the above example, that the method of computing the mean place of a planet by this system, is less troublesome by far in the operation, than in the system of Brahma. If we wanted to find the mean longitude of Venus for the end of the year 4029 of the *Kali Yuga*, by the latter system, we must find the time elapsed from the creation, thus:—

To sunrise at the beginning of the *Kali Yuga*, it is . .	1972944000
Add, ...	4029
To the end of the year 4029 of the *Kali Yuga*, ...	1972948029

The number of revolutions of Venus in 4320000000 years is 7022389492: therefore the mean heliocentric longitude of Venus at the end of the year 4029 of the *Kali Yuga*, by this system, will be expressed by the formula: $\frac{7022389492 \times 1972948029}{4320000000} = \frac{13854809507111711268}{4320000000} = 3207131830$ revolutions, $4^s\ 5^\circ\ 58'\ 33''\ 23'''$, the mean longitude of Venus at the end of the year 4029 of the *Kali Yuga*. This shows not only the great labour in the calculation, but also partly the error in position: to get the whole error we must compare the time; for the end of the year 4029 of the *Kali Yuga*, by the system of Brahma, does not coincide with the end of the same year by the system of Varāha. For, 4029 years, by the system of Brahma, $= 365^{ds.}\ 15^{da.}\ 30'\ 22''\ 30''' \times 4029 = 1471626^{ds.}\ 14^{da.}\ 40'\ 52''\ 30'''$ and 4029 years, by the system of Varāha, $= 1471627^{ds.}\ 31^{da.}\ 47'\ 30''\ 36'''$ the difference is, $1^{d.}\ 17^{da.}\ 6'\ 38''\ 6'''$ but the former begins from sunrise, and the latter from midnight: therefore we must diminish the difference by six hours, or 15 dandas, which will make $1^{d.}\ 2^{da.}\ 6'\ 38''\ 6''' = 1^{d.}\ 0^{h.}\ 50^{m.}\ 39''\ 14'''\ 24^{iv.}$ the real difference in A.D. 928. We must therefore add the mean motion of Venus for this difference $= 1^\circ\ 39'\ 30''$ to $4^s\ 5^\circ\ 58'\ 33''\ 23'''$ the sum is $4^s\ 7^\circ\ 38'\ 3''\ 23'''$ the mean longitude of Venus by the system of Brahma in A.D. 928, at the moment the longitude by the system of Varāha was $4^s\ 2^\circ\ 44'\ 31''\ 12'''$ which, as the latter only differed $5'\ 7''\ 48'''$ from the truth, shows how erroneous the system of Brahma had become at that time; though in A.D. 538, when it was framed, it gave the place of Venus to within 6′ 41″ of the truth. This is, in fact, the case with every system: they are all correct, or nearly so when framed, but not so at any considerable distance of time, either before or after. The mean longitude of Venus, by the system of Varāha at the end of the year 4029 of the *Kali Yuga*, was found above not to differ 6′ from the truth; but will it give the position of that planet in A.D. 538, or the end of the year 3639 of the *Kali Yuga*, with the same degree of correctness? Most certainly not, because it was not then framed. The mean longitude of Venus at the end of the year 3639 of the *Kali Yuga*, by the system of Varāha =

$\frac{1755594 \times 3639}{1080000} =$ . .	4^s	15°	31	$19''$	$12'''$
By LaLande's Tables for the same instant, H. sphere,	4	18	46	3	44
Error in A.D. 538,	0	3	14	44	32

And if we carried our calculations still farther back into antiquity, the error would be found to increase in proportion to the time, so that at the beginning of the *Kali Yuga* the error would amount to 32° 43′ 36″ Upon this change in the error in proportion to the time, is founded the method of determining the antiquity of astronomical systems or books; and, in fact, it is not only the surest, but the very best that can be employed. If there were no errors, or if the errors were always the same, then we should have no data to proceed on; but this supposition is in its nature impossible: there never yet was found any set of astronomical tables or systems, whether European or otherwise, that did not in progress of time become more and more inaccurate, by the continual accumulation of the errors in the mean annual motions.

We shall now proceed to show the antiquity of the system of Varāha, in the same manner as we did that of Brahma, by dividing the errors in position at the beginning of the *Kali Yuga* by the errors in the mean annual motions.

The following Table shows the positions of the planets, both by the system and La Lande's Tables, at the instant of midnight, at the beginning of the *Kali Yuga*, on the meridian of *Ujein*, 75° 50′ east of Greenwich.

THE TABLE.

	By La Lande, H. Sphere.				By the System.				Errors or Differences.			
Sun, ...	0s	0°	0′	0″	.. 0s	0°	0′	0″	0s	0°	0′	0″
Moon, ...	0	0	5	56	.. 0	0	0	0	− 0	0	5	56
—— Apogee,	4	0	11	25	.. 3	0	0	0	− 1	0	11	25
—— Node,	5	6	22	29	.. 6	0	0	0	+ 0	23	27	31
Mercury, ...	10	26	34	25	.. 0	0	0	0	+ 1	3	25	35
Venus, ...	1	2	43	36	.. 0	0	0	0	− 1	2	43	36
Mars, ...	11	17	54	18	.. 0	0	0	0	+ 0	12	5	42
Jupiter, ...	0	17	2	53	.. 0	0	0	0	− 0	17	2	53
Saturn, ...	11	9	0	57	.. 0	0	0	0	+ 0	20	59	3

And the following Table exhibits the mean annual motions of the planets, both by the system and La Lande's Tables, with the differences or errors.

Mean Annual Motions of the Planets, &c. Hindu Sphere.

	By La Lande's Tables.				By the System.				Differences.
Sun, ...	12s	0°	0′	0″	.. 12s	0°	0′	0″	
Moon, ...	4	12	46	40 ,613	.. 4	12	46	40 ,8	.... + 0″,187
—— Apogee,	1	10	40	35 ,591	.. 1	10	41	00 ,9	.... + 25 ,309
—— Node,	0	19	21	31 ,090	.. 0	19	21	11 ,4	.... − 19 ,690
Mercury, ..	1	24	45	36 ,943	.. 1	24	45	7 ,2	.... − 29 ,743
Venus, ...	7	15	11	23 ,635	.. 7	15	11	52 ,8	.... + 29 ,165
Mars, ...	6	11	24	19 ,15	.. 6	11	24	9 ,6	.... − 9 ,55
Jupiter, ...	1	0	20	50 ,483	.. 1	0	21	6 ,0	.... + 15 ,517
Saturn, ...	0	12	13	9 ,343	.. 0	12	12	50 ,4	.... − 18 ,949

Dividing the errors in position at the beginning of the *Kali Yuga*, by the differences in the mean annual motions, we have from

						Years.
The Moon, ..	5′ 56″ + (7′ 26″ the Sec. Eq.)			divided by	0″,187	=4288
—— Apogee,	30°	11′	25″	by	25 ,309	=4294
—— Node, ..	23	27	31	by	19 ,690	=4289
Mercury, ..	33	25	35	by	29 ,743	=4046
Venus, ...	32	43	36	by	29 ,165	=4040
Mars, ...	12	5	42	by	9 ,55	=4559
Jupiter, ...	17	2	53	by	15 ,517	=4032
Saturn, ...	20	59	3	by	18 ,943	=3988

The sum is 33536

Which divided by 8, gives for a mean result, ... 4192

Or the year A.D. 1091: whence it appears that the system must have been framed a good many years after the observation attributed to *Vridha Vasishtha* on the star Canopus, above mentioned.

We shall now exhibit the errors in the positions of the planets, &c. by the system, compared with La Lande's Tables, at different periods, from the beginning of the *Kali Yuga* down to the year 4192, or A.D. 1091 in the Table following:

Planets, &c.	Kali Yuga.	Kali Yuga 1000.	Kali Yuga 2000.	Kali Yuga 3000.	K. Y. 3639, 538 A. D.	K. Y. 4192. 1091 A.D.
Moon,*	− 5°52′41″	− 3°50′48″	− 2° 9′17″	−0°52′33″	−0°18′30″	−0° 0′11″
Moon's apogee	−30 11 25	−23 9 36	−16 7 47	−9 5 58	−4 36 26	−0 43 10
Moon's Node	+23 27 31	+17 59 21	+12 31 11	+7 3 1	+3 33 19	+0 31 50
Mercury,	+33 25 35	+25 9 52	+16 54 9	+8 38 26	+3 21 40	−1 12 28
Venus,	−32 43 36	−24 37 31	−16 31 26	−8 25 21	−3 14 45	+1 14 3
Mars,	+12 5 42	+ 9 26 32	+ 6 47 22	+4 8 12	+2 26 30	+0 58 29
Jupiter,	−17 2 53	−12 44 16	− 8 25 39	−4 7 2	−1 21 47	+0 41 14
Saturn,	+20 59 3	+15 43 20	+10 27 37	+5 11 54	+1 50 10	−1 4 25

The above Table serves to show, by mere inspection, the time at, or near which the system was framed, by the gradual decrease in the errors down to the year A.D 1091, after which they would again increase.

Having already given the Lunar Asterisms in treating of the system of Brahma, it might be considered as altogether unnecessary to repeat them over again

* Including the secular equation.

here from the system of Varāha, since from their very nature being sideral and consequently fixed, they must be the same, or nearly so, by all Hindu writers, whether ancient or modern, except where errors may have crept in. But the author of the *Surya Siddhanta* having noticed in that work, certain deviations in the positions of some of the stars, by which they appear to fall into other Lunar Asterisms, different in name from those to which they originally did belong, I am, therefore, induced to give them here a place, for the purpose of explaining the cause of the deviation alluded to.

Table of the Lunar Asterisms, according to the Surya Siddhanta.

Names.	Latitudes.*	Longitudes from Aswinī.*	Longitude in the Mansion.	Stars supposed to be intended.
1 *Aswinī*, ..	10° N.	8° 0′	8° 0′	γ or β Arietis.
2 *Bharanī*, ..	12	20 0	6 40	36 ditto.
3 *Kṛitikā*, ..	5	37 30	10 50	Alcyone?
4 *Rohinī*, ..	5 S.	49 30	9 30	87 Tauri.
5 *Mṛigasiras*, ..	10	63 0	9 40	113, 116, 117 Tauri?
6 *Ardrā*, ..	9	67 20	0 40	133 Tauri?
7 *Punarvasu*, ..	6 N.	93 0	13 0	β Geminorum.
8 *Pushyā*, ..	0	106 0	12 40	δ Cancri.
9 *Asleshā*, ..	7 S.	109 0	2 20	49, 50 Cancri.
10 *Maghā*, ..	0 N.	129 0	9 0	Cor Leonis.
11 *P. Phalgunī*, ..	12	144 0	10 40	70, 71 Leonis.
12 *U. Phalgunī*, ..	13	155 0	8 20	β Leonis.'
13 *Hastā*, ..	11 S.	170 0	10 0	7, 8 Corvi.
14 *Chitrā*, ..	2	180 0	6 40	Spica Virginis.
15 *Swāti*, ..	37 N.	199 0	12 20	Arcturus.
16 *Visākha*, ..	1 30 S.	213 0	13 0	24 Libræ.
17 *Anurādhā*, ..	3	224 0	10 40	β Scorpii.
18 *Jyesht'ha*, ..	4	229 0	2 20	Antares.
19 *Mulā*, ..	9	241 0	1 0	34, 35 Scorpii.
20 *P. Āshād'ha*, ..	5 30	254 0	0 40	δ Sagittarii.
21 *U. Āshād'ha*, a ..	5	260 0	6 40	φ Sagittarii.
* *Abhijit*, b ..	60 N.	266 40	13 20	α Lyræ.
22 *Sravanā*, c ..	30	280 0	13 20	α Aquillæ.
23 *Dhanisht'ha*, d ..	36	290 0	10 0	α Delphini.
24 *Satabhishā*, ..	0 30 S.	320 0	13 20	λ Aquarii.
25 *P. Bhādrapadā*,	24 N.	326 0	6 0	α Pegasi.
26 *U. Bhādrapadā*,	26	337 0	3 40	γ Pegasi.
27 *Revatī*, ..	0	359 50	13 10	ζ Piscium.

* What are called latitudes and longitudes in this Table, are only the distances already explained at page 99 by the Diagram.

By comparing the above Table with the one given under the system of Brahma, it will appear, that in general they agree, or at least nearly so: they differ, however, in one or two instances, very materially, which must be attributed to errors having crept in by miscopying. In the table of the *Brahma Siddhanta*, the star *Purva Phalgunī*, θ Leonis, stands in 147° from *Aswinī*; but by the *Surya Siddhanta*, it should be 144°: the latter is nearer the truth. Spica Virginis, *Chitrā*, is placed by the *Brahmā Siddhanta* in 6^s 3°, or 183° from *Aswinī*; but the *Surya Siddhanta* places it in 6^s, or 180°, making a difference of three degrees. In this respect, the *Surya Siddhanta* is in error. For, taking the position of Cor Leonis as correctly given, it being the same by all Hindu books, then Spica should be in 6^s 3°: for the difference in longitude between Cor Leonis and Spica, by European books, is about 54°; which being added to the longitude of Cor Leonis, 129°, we get 183°, the longitude of Spica, agreeing with the *Brahma Siddhanta*. A few other differences may be observed, but they are of less importance.

Now with respect to the passage above alluded to, the author of the *Surya Siddhanta* states, that the star *Uttara Āshād'hā*, (No. 21,) falls into the middle of *Purva Āshād'hā*, (No. 20); that the star *Abhijit*, (marked in the table with an *,) falls into the end of *Purva Āshād'hā*, (No. 20); that the star *Sravanā*, (No. 22,) falls into the end of *Uttara Āshād'hā*, (No. 21); and that the star *Dhanisht'hā*, (No. 23,) is in the end of the third quarter of *Sravanā*, (No. 22). All this, though strange in appearance, is very true, and easily accounted for. The star *Abhijit* is given as the key to the mystery; for it does not belong to the division of the zodiac into 27 parts, and could not, therefore, fall into any Lunar Asterism of its own name. At the commencement of the Hindu astronomy, the zodiac was divided into 28 equal parts, each containing 12° 51′$\frac{3}{7}$, and the first of such divisions was called *Mulā*, thereby to signify that it was the root or origin in the series. This division of the zodiac, was, however, found to be rather inconvenient in the practice of astronomy: therefore it was changed to 27 equal portions, each containing 13° 20′. The first of these divisions was called *Jyesht'hā*, to denote that it was the first or eldest in the series, and began from the same point in the heavens as *Mulā* in the division of 28. The star that belonged to *Mulā* was Antares, and its longitude was about 2° 25′ from the beginning of that asterism, and consequently had the same longitude in the Lunar Asterism *Jyesht'hā*. In this new arrangement, *Abhijit* was thrown out; but the names of all the rest were retained, though not to the same stars that originally belonged to them. The name *Mulā* was given to the second, or next mansion to *Jyesht'hā*, and other changes made to answer the arrangement. The following short Table will explain the cause of the deviations: —

Division of Twenty-eight Mansions of 12° 51′ 3-7 each.		Division of Twenty-seven Mansions of 13° 20′ each.		Stars supposed to be intended.
Names.	Longitude.	Names.	Longitude.	
1 *Mulā*	2° 25′	1 *Jyesht'hā*,	2° 25′	Antares.
2 *P. Āshād'hā*,	1 28 4–7	2 *Niriti*	1 0	δ Sagittarii.
3 *U. Āshād'hā*,	1 37 1–7	3 *P. Āshād'hā*,	0 40	
×3 *U. Āshād'hā*,	7 37 1–7	3 *P. Āshād'hā*,	6 40	φ Sagittarii.
×4 *Abhijit*,	1 25 5–7	3	13 20	α Lyræ.
×5 *Sravanā*,	1 54 2–7	4 *U. Āshād'ha*,	13 20	α Aquilæ.
×6 *Danisht'hā*,	0 0	5 *Sravanā*,	10 57	α Delphini.

This Table, as far as it has been thought necessary to carry it, contains the corresponding positions of the stars in both divisions. Thus the longitude of the star in *Mulā* 2° 25′ in the division of 28, is the same in *Jyesht'hā* in the division of 27. The longitude 1° 28′ 4-7 in *P. Āshād'hā,* in the division of 28, becomes 1° in *Niriti.* The star *U. Āshād'hā,* whose longitude is 7° 37′ 1-7 in the division of 28, falls into 6° 40′ of *P. Āshād'hā,* in the division of 27. The star *Abhijit,* whose longitude is 1° 25′ 5-7 in the mansion *Abhijit,* in the division of 28, falls into the end, or 13° 20′ of *P. Āshūd'ha* of the division of 27. The longitude of the star *Sravanā,* 1° 54′ 2-7 in the division of 28, falls into the end, or 13° 20′ of *U. Āshād'hā;* and the star in the beginning of *Danisht'hā* in the division of 28, falls into 10° 57′ of *Sravanā* in the division of 27: the whole of which corresponding to what the author of the *Surya Siddhanta* states, and shewing, in a clear manner, that the cause of the supposed deviations, arises from still using the names of three Lunar Asterisms, which belong to the division of 28, without its being known or suspected that they are so. Mr. Colebrooke says, the cause arises from the longitudes being reckoned by the circles of declinations, and not by the circles of latitude, cutting the ecliptic; but this circumstance could not cause the deviations alluded to. We shall now take a view of the subject of the precession of the equinoxes. The method given in the system of Varaha (contained in the *Surya Siddhanta*), for computing the precession, differing widely from that of the system of Brahma, a particular explanation of the cause and foundation of that method may perhaps not only be acceptable, but also useful in doing away the incorrect notions that have been entertained by some on that subject.

By the system of Brahma, the number of revolutions of the equinoxes in a *Kalpa*, or 4320000000 years=199669. These revolutions are retrograde, and were determined in the same manner as those of the planets already explained. Now, suppose we wanted to determine the precession of the equinoxes for the end of the year 4900 of the *Kali Yuga* (A.D. 1799), from this number, we must first find the years elapsed of the system, thus:—

To the beginning of the *Kali Yuga*,	1972944000
Add, ...	4900
Total years to April 1799, .	1972948900

Then we get the precession by the following formula, viz.

$\frac{199669 \times 1972948900}{4320000000} = \frac{393936733914100}{4320000000} = 91189^{rev.}\ 0^{s}\ 21^{\circ}\ 9'\ 34''.205$; that is to say, the precession of the equinoxes was then (April, A.D. 1799), 21° 9′ 35″.205, or the quantity by which the vernal equinoctial point had fallen back from the beginning of *Aswini*. This example is sufficient to show how troublesome the operation of finding the precession is, from the number given in the system of Brahma. So it must have likewise appeared to the author of the system of Varāha: but how to remedy the evil was a task of no small difficulty. He succeeded in lessening the labour of calculating the places of the planets, by giving their revolutions in small cycles: but here that method could not answer, because the period of one single revolution of the equinoxes, would exceed 25000 years; and to begin such periods from the commencement of the *Kalpa*, they would become equally as troublesome in computation, as the number in the system of Brahma. Therefore, to avoid all this, he conceived, that as the most perfect astronomical system that was ever framed could not always last, the best plan would be, to make his rule answer within a certain limited period of time. He was aware, that the earliest observation the Hindus had on record, only placed the vernal equinoctial point in the beginning of *Criticā*, or 26° 40′ to the east of the beginning of *Aswinī*: therefore, by taking into his rule, as far as 27° on the east side, and just as many on the west side, he would not only include the most ancient observations, but also give a sufficient scope of time to elapse before his rule would become useless. The next thing was to adjust this space of 27° on each side of the beginning of *Aswinī*, to time and circular motion; for without the idea of circular motion, he could not connect it with his system. He therefore assumed the space of time in which the equinoxes would fall back 54°, or 27° × 2, at 3600 years. Then, to get a circular motion, he assumed the equinoxes to move in the periphery of an epicycle, the centre of which is fixed to the beginning of *Aswinī*, and the dimensions of the periphery 108°, or 54° × 2, so that one complete revolution of the equinoxes in the epicycle would be 7200 years. By this ingenious contrivance, he transfers the 54° in the zodiac to the periphery of the epicycle, of which it takes up the lower half. The rest of the contrivance will now be explained by the following Diagram: —

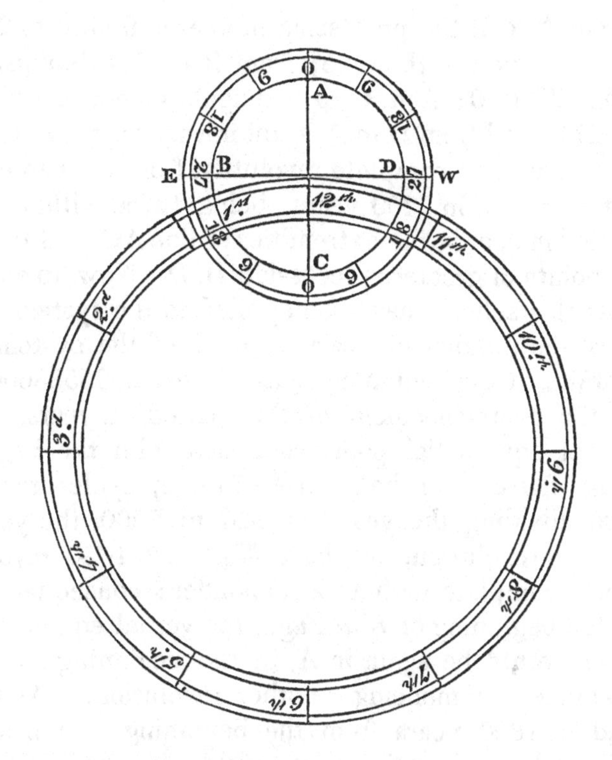

Let the large circle be the Hindu zodiac, divided into 12 parts or signs, and marked 1st, 2nd, 3rd, &c.

And ABCD the epicycle, the centre of which is at 12 in the beginning of *Aswinī*.

The line AC, divides the epicycle into two equal halves; and as it cuts through the beginning of *Aswinī* at 12, its two extremities in the periphery of the epicycle form the points of superior and inferior conjunctions, where the precession is nothing. In the point B the precession is at its greatest quantity to the east, viz. 27°, and in the point D it is at its greatest quantity to the west, or 27°, these being the limits beyond which it does not increase by the rule.

From A to B the precession increases from 0 to 27°, in proportion to the time; from B to C it diminishes from 27° to 0; from C to D it again increases from 0 to 27°; and from D to A it diminishes from 27° to 0; thus making a complete revolution from west to east through 108° in 7200 years, being twice within that period in 0, or in the extremities of line AC, and twice in points of greatest precession B, D. Now to show how this scheme has been applied to the system, we have the number of years elapsed of the system of Varāha, at the beginning of *Kali Yuga*=1955880000: at the commencement of this period of years, the vernal equinoctial point coincided with the beginning of *Aswinī*, in the point A of the epicycle: therefore, dividing the years elapsed by 7200, the years in one revolution, we have $\frac{1955880000}{7200}=271650$ revolutions, complete without a remainder: consequently, at the beginning of *Kali Yuga*, the vernal equinoctial point would be again in A, in the beginning of the epicycle commencing another revolution. At the end of 1800 years from the beginning of the *Kali Yuga*, the equinoctial point would arrive at B, where its distance from *Aswinī*, reckoned on the periphery of the epicycle, would be 27° to the east. The vernal equinoctial point, still moving onwards in the periphery, diminishes its distance from *Aswinī*, until in the year 3600 of the *Kali Yuga*, A.D. 499, it coincided with the point C or 0, being then in the line with the beginning of *Aswinī*; and therefore in that year the precession was 0. Since that time, the equinoctial point proceeds from C towards D; and therefore the precession must continually increase, until it amounts to 27° at D; after which (by the scheme), it proceeds from D to A, and completes the revolution.

From the explanation thus given, it must be easy to perceive how the precession is to be computed. In 1800 years, the precession increases to its greatest quantity, 27°. Therefore all that we have to do, is to say, as 1800 years, to 27°, so any number of years less than 1800, to the corresponding precession; which is to be counted to the east of *Aswinī,* in the two first quadrants of the epicycle, but to the west in the two last.

Thus, suppose we wanted to know the precession of the equinoxes, or the distance of the vernal equinoctial point from the beginning of *Aswinī,* for the end of the year 4900 of the *Kali Yuga,* (A. D. 1799,) we subtract two periods of 1800 years each=3600, the remainder is 1300 years; then 1800: 27°:: 1300: 19° 30, which is to the west of the beginning of *Aswinī,* being in the third period of the cycle. This example is given for the same year, that the calculation was made for by the system of Brahma, in order to show the difference in labour, &c.: by the latter, the precession was found to be 21° 9′ 34″.2, which is nearest the truth.

I have been more particular in my explanation of the contrivance of the author of the system of Varāha for calculating the precession of the equinoxes, than perhaps was necessary; but my reason for it was to do away an erroneous notion that appeared to be entertained by some who called the motion a libration*, or oscillation of the equinoxes, instead of a complete revolution, which the author himself expressly mentions; for he says: "The *Ayanānsa* moves *eastward* thirty times twenty=(600) in each *Maha Yuga:*" therefore each revolution $=\frac{4320000}{600}=7200$

* As. Res. Vol. xii. p. 212, 217, 218.

years, which, therefore, must be conceived to be in an epicycle, as described above. The *Sācalya Sanhita* states, "that the *Bhaganas*, (revolutions,) of the *Crāntipāta*, (point of intersection of the ecliptic and equator,) in a *Maha Yuga*, (4320000 years,) are 600 eastward," in which I see not the slightest shadow for conceiving the idea of an oscillation, or libration, at least not according to my conception of the terms. The commentator on the *Surya Siddhanta* is still more explicit. He says: "The *Bhaganas*, (revolutions,) of the *Ayanānsa*, (equinoctial points,) in a *Maha Yuga*, (4320000 years,) are 600; one *Bhagana*, (or revolution) of the *Ayanānsa*, therefore, contains 7200 years. He then describes how the revolution is divided, thus: "Of a *Bhagana*, (revolution,) there are four *pādas*, quadrants, or parts. First *Pāda*, when there was no *Ayanānsa*," (as when the vernal equinoctial point was at C in the epicycle, in A.D. 499;) "but the *Ayanānsa*, (precession,) beginning from that time, and increasing (from C to D,) it was added. It continued increasing 1800 years, when it became at its utmost, or 27°, (as at D in the epicyle.) Second *Pāda*, (or quadrant) after this it diminished; but the amount was still added, (because to the west of *Aswinī*,) until the end of 1800 years more, it was diminished to nothing, (as at A in the epicycle.) Third *Pāda*, the *Ayanānsa* for the next 1800 years was deducted; (that is from A to B, because to the east of *Aswinī*;) and the amount deducted at the end of that term was twenty-seven degrees. Fourth *Pāda*, (from B. to C,) the amount of deduction diminished; and at the end of the next term of 1800 years, there was nothing either added or subtracted," because it had then returned to C, where the precession was nothing, being in a line with the beginning of *Aswinī*. The commentator, however, has made a mistake; for the first *Pāda*, or quadrant of the revolution did not begin at C, but at A, for this reason, that the rule was intended to give the precession from B to C, and from C to D, the other two quadrants being fictitious, and added merely to introduce into the calculation a circular motion: and as the quadrant B C was prior in point of time, to C D, it must follow, that CD could not be the first quadrant, consistently either with calculation or the nature of the scheme. But in the system of Āryabhatta, which the commentator seems to have followed, the first quadrant begins at C, which may have occasioned the mistake.

In all this contrivance, I see nothing that could support in the slightest degree the opinion that has been formed by some, that the author of the *Surya Siddhanta* believed in a libration of the equinoxes. We may suppose any thing, if we do not chuse to give ourselves the trouble of investigating, and fairly entering into the author's ideas and intentions, which we should always hold in view. An erroneous opinion must be always the consequence, when the matter before us is not properly understood; of which we have repeated and most decisive proofs, in various instances.

SECTION III.

The Ārya Siddhanta—by Āryabhatta—Its date, A. D. 1322—The object of it—The system it contains—Its formation—Precession of the equinoxes—Mode of computing it—The Rishis—The object of introducing them, and the manner of computing their place—The Pārāsara Siddhanta, and the object of the author in exhibiting it—The system of the Pārāsara Siddhanta—The computation of the Rishis by this system—Age of the Ārya Siddhanta, confirmed by computations from Astronomical Data—Age of the system of Pārāsara determined—Found to be of the same age with that of Āryabhatta—Latitudes and Longitudes of the Stars—Geometry of Āryabhatta same as Bhāskara's—His rules for shewing the proportion of the diameter of a circle to its circumference, &c.

THE next astronomical work in point of antiquity, that I have met with, is the *Ārya Siddhanta,* written by Āryabhatta, in the year 4423 of the *Kali Yuga,* or A.D. 1322. It is divided into eighteen chapters or sections, in the first of which he gives both the date and the system. His principal objects appear to have been, first, to give a system that would give the position of the planets, agreeing with their real places in the heavens, in his own time, much nearer the truth than could be then obtained from either the *Brahma Siddhanta,* the *Surya Siddhanta,* or any other work then extant. Secondly, to support the modern impositions respecting the introduction of immense periods of years into their history; and, lastly, to endeavour, by a curious contrivance, to pervert the meaning of the passage I have mentioned and explained in a former part of this essay, respecting the *Rishis* being in *Maghā* in the time of Yudhist'hira,

Pārāsara, &c. Before, however, we can enter on this discussion, it will be proper to give Āryabhatta's system entire in the following Table:

ĀRYABHATTA'S SYSTEM,

FROM THE FIRST CHAPTER OF THE ĀRYA SIDDHANTA.

Revolutions of the Planets, Apsides, and Nodes in 4320000000 *Years.*

Names.	Revolutions.	Apsides.	Nodes.
Sun, ...	.. 4320000000	Revol. 461	
Moon, ...	.. 57753334000	.. 488108674	Revol. 232313354
Mercury, ...	.. 17937054671	.. 339	.. 524
Venus, ...	.. 7022371432	.. 658	.. 947
Mars, ...	.. 2296831000	.. 299	.. 298
Jupiter, ...	.. 364219682	.. 830	.. 96
Saturn, ...	.. 146569000	.. 36	.. 620

Revolutions of the *Rishis* in a *Kalpa*, ...	1599998
Revolutions of the equinoxes in the epicycle in ditto,	578159
Solar months .... in ditto,	51840000000
Lunations .... in ditto,	53433334000
Intercalary months .... in ditto,	1593334000
Tithis, or lunar days .... in ditto,	1603000020000
Intercalary tithis .... in ditto,	25082478000
Sideral days .... in ditto,	1582237542000
Natural days .... in ditto,	1577917542000

The system of Āryabhatta begins on Sunday, as the first day of the week, on the meridian of *Ujein*, at sunrise: and the number of years elapsed of the system, is known by deducting 3024000, from the years elapsed of the system of Brahma. Thus, by the system of Brahma, the years elapsed at the beginning of the

Kali Yuga were,..................................	1972944000
Deduct, ...	3024000
Remain years elapsed, by the system of Āryabhatta, ...	1969920000

The length of the year $=\frac{1577917542000}{4320000000}=365^{ds}\ 15^{da}\ 31'\ 17''\ 6'''$

The system of Āryabhatta, was constructed precisely on thr same principle as that of Brahma:

therefore no further explanation is necessary here, except, that the differences between the numbers arise from the systems being framed for two distinct periods of time; that of Brahma, for the year A. D. 538, and that of Āryabhatta, for the year A. D. 1322; for it may be easily conceived, that the numbers that will answer at one particular period of time, will not answer at another.

The computation of the precession of the equinoxes, by the system of Aryabhatta, is on a similar plan with that given in the *Surya Siddhanta*, already explained; but by no means so convenient in practice. Āryabhatta gives 578159 revolutions of the equinoxes, in the epicycle, in 4320000000 years; which makes one revolution equal to 7471.993 years, or say, for common practice, 7472 years, whereas the *Surya Siddhanta* makes it 7200 years. The number of revolutions of the equinoxes in the epicycle, to the beginning of the *Kali Yuga*, by the system of Āryabhatta $= \frac{1969920000 \times 578159}{4320000000} = 26364$ revolutions, and .504, or a little more than half a revolution over and above. This .504, being the only part we want, might be got much easier by multiplying the three last figures of 578159 by 456, reserving only the three right hand figures in the product, thus, $159 \times 456 = .504$. Now, as one revolution, is to 7472 years, so the fraction .504 to 3765.888 years; or the number of years elapsed at the beginning of the *Kali Yuga* since the vernal equinoctial point was in the beginning of the epicycle: subtract half a period, or 3736, the remainder 29.888, or near 30 years, was the time anterior to the beginning of *Kali Yuga*, when the equinoctial point was in the middle of the epicycle. The dimensions of the epicycle in the system of Āryabhatta is 96°, that is to say, each quadrant 24°; but in the *Surya Siddhanta,* each quadrant is 27°: in other respects, the explanation is the same. From a fourth part of the entire revolution=1868 years, subtract 29.888, we get 1838.112, the year of the *Kali Yuga,* when the precession was 24° eastward by this system; and as it decreased from that time, if we add one fourth of a revolution, or 1868 years, we get 3706.112, the year of the *Kali Yuga,* when the precession was nothing, and the vernal equinoctial point in the beginning of the epicycle, in a line with the beginning of the Lunar Asterism *Aswinī.* This corresponded to the year A. D. 605. From these data, it is easy to determine the precession at any time since the year 3706, by the rule of proportion, by saying, as 1868 years to 24 degrees, so the number of years elapsed since the year 3706 to the precession. Thus, suppose required for the year of the *Kali Yuga* 4900, (A. D. 1799,) subtract 3706, the remainder is 1914; then as 1868: 24°:: 1194: 15° 20′ 25″, the precession, or distance of the vernal equinoctial point from the Lunar Asterism *Aswinī,* reckoning on the periphery of the epicycle to the west. This quantity is by far too small, owing to the erroneous rate of precession made use of by Āryabhatta, which he makes only 46″.2526 per annum. The precession for the year 4423 of the *Kali Yuga,* (A. D. 1322,) when Āryabhatta wrote, would, by the same rule, be 9° 12′ 43″.

Having thus explained the precession according to the scheme of Āryabhatta, we shall now proceed to show his contrivance for doing away, or perverting the original meaning of the passage already alluded to, respecting the *Rishis* being in *Maghā* in the time of Yudhist'hira, Pārāsara, &c. which term

I have already explained in pp. 64, 65, 66, and simply related to the precession of the equinoxes, and the motions of the Lunar Mansions depending on them, reckoning from the year 1192 B.C. when there was no precession, the fixed and moveable Lunar Mansions of the same name, then coinciding. Āryabhatta, finding this passage an obstacle in the way of transferring the ancient history of the Hindus to the immense periods of the modern system of Brahma, thought he might be able to remove the difficulty, by giving it a different explanation and turn; which would at once, with all the astronomers and Brahmins, immortalize his name for ever. For this purpose, he assumed the *Rishis*, (the seven stars in the Great Bear,) to have a particular motion of themselves, different from all the rest of the stars, by which they moved, or were feigned to move eastward, at the rate of 13° 20′, or one Lunar Mansion every hundred years, or nearly so; thereby making a complete revolution in about 2700 years, or rather 1599998 revolutions in a *Kalpa*, or 4320000000 years as given by himself in the system.

At the commencement of his *Kalpa*, (1969920000 years before the beginning of the *Kali Yuga*,) he assumes the line of the *Rishis* to be in the beginning of *Aswinī*. Therefore, to find the position of that line at the beginning of the *Kali Yuga*, multiply the three right hand digits of the number 1599998 by 456, reserving the three right hand figures in the product, and we have $998 \times 456 = .088$, the decimal parts of a revolution at the beginning of the *Kali Yuga;* to find the value of which in years, say, as $1^{rev.} : 2700^{yrs.} :: .088^{rev.} : 237.6^{yrs.}$ which shows, that at the beginning of the *Kali Yuga*, 237 years and 6-10ths had elapsed from the time the *Rishis* were in *Aswinī;* and as the *Rishis* were feigned to move one Lunar Mansion to the east, every hundred years, the mansion in which the line of the *Rishis* then fell into, was *Kriticā,* 37 years. Now to find the time of Pārāsara, &c. we must add 100 years for each Lunar Mansion, till we come down to *Maghā,* thus:—

Remainder of *Kriticā,*	. . years	62.4
Add for *Rohinī,*	. . .	100
for *Mṙigasiras,*	. . .	100
for *Ardrā,*	. . .	100
for *Punarvasu,*	. . .	100
for *Pushyā,*	. . .	100
for *Asleshā,*	. . .	100
Total to the beginning of *Maghā,*		662.4

So that by this silly contrivance, the *Rishis* were feigned to be in the beginning of *Maghā,* in the 663d year of the *Kali Yuga,* or B.C. 2439; and Pārāsara, Yudhishthira, &c. who lived between five and six hundred years before Christ, were thrown back into antiquity upwards of 1800 years; nay, upwards of 4500 years, if he meant to place them in the *Dwāpar Yuga;* for, in that case, we must go back one full period of 2700 years more. But neither of these agree even with the modern system, and not at all with the ancient method. It is natural, therefore, to suppose that Āryabhatta's imposition would not gain credit, but, on the contrary, be opposed. For, Āryabhatta's contemporaries would naturally ask, where did he get all this supposed knowledge? for he could know no more on matters of antiquity than they themselves did; and, moreover, that the *Rishis* had no other motions in the heavens, but what all the stars had in common. To silence all such impertinent questions, and to put all cavils to an end, he conceived that the best plan would be to follow up the imposition by another of a more formal nature, and against which, the same arguments that might be used against himself as a modern, could not so effectually apply. This was nothing more nor less than to forge and construct another work and father it on Pārāsara, giving it the name of the *Pārāsara Siddhanta.* By this second piece of imposition, he could reply, that whether the *Rishis* had a separate motion of their own, differing from all the rest of the stars, or not, was a matter that had nothing to do with him, or whether he believed in such motion or not; that the ancients had employed it for the purpose of computation, and as a proof of it, that it still existed in the genuine work of Pārāsara, entitled the *Pārāsara Siddhanta;* and that if they did not believe that work, they would not believe Pārāsara himself, if he rose from the dead to confirm it. Such, I conceive, would naturally be the arguments that would be employed by Āryabhatta to support his impositions. Having thus far explained the matter, we shall now give the system of Pārāsara, from the second chapter of the *Ārya Siddhanta.*

SYSTEM OF PĀRĀSARA.

Revolutions of the Planets, &c. in a Kalpa, or 4320000000 Years, according to the Pārāsara Siddhanta.

Planets.	Revolutions.	Apsides.	Nodes.
Sun, ...	.. 4320000000	.. 480	
Moon,............	.. 57753334114	.. 488104634	.. 232313235
Mercury, ...	.. 17937055474	.. 356	.. 648
Venus, ...	.. 7022372148	.. 526	.. 893
Mars,............	.. 2296833037	.. 327	.. 245
Jupiter, ...	.. 364219954	.. 982	.. 190
Saturn, ...	.. 146571813	.. 54	.. 630

Revolutions of the seven *Rishis* in 4320000000 years =	1599998
Revolutions of the equinoxes in the epicycle in ditto,	581709
Solar months .... in ditto,	51840000000
Lunations .... in ditto,	53433334515
Intercalary ditto .... in ditto,	1593334515
Tithis, or lunar days .... in ditto,	1603000035450
Natural days .... in ditto,	1577917570000
Intercalary lunar days.......................... in ditto,	25082465450
Sideral days .... in ditto,	1582237570000

This system was constructed precisely in the same manner as the system of Brahma, and therefore requires no further explanation on that account here. The number of years elapsed of the system at the beginning of the *Kali Yuga*, is the same as in the system of Brahma, viz. 1972944000, and the days elapsed to the same epoch were 719530411920. The *Kalpa* also begins on Sunday at sunrise, on the meridian of *Ujein*, where in fact the system was framed.

From the circumstance of the years elapsed at the beginning of the *Kali Yuga*, being the same as in the system of Brahma, a person unacquainted with the structure of these systems, would wonder why the revolutions of the planets, apsides, and nodes, are not the same in both. The reason is this, the system of Brahma was constructed to give the positions of the planets when its author lived, that is, about A.D. 538. The system of Pārāsara, for the same reason, was constructed to give the positions of the planets in the time of its inventor, that is to say, in the time of Āryabhatta, or about A. D. 1322. For the same number of revolutions of the planets will not give their positions true at two distant periods of time; and hence arises the necessity of forming new systems, from time to time.

The length of the year, by the system of *Pārāsara*, =
$\frac{1577917570000}{4320000000}$ = ... 365^{ds} 15° $31'$ $18''$ $30'''$
By the *Ārya Siddhanta*, it is 365 15 31 17 6
The time of one revolution of the equinox in the epicycle, by the system of *Pārāsara*, ...=7426.3936
By the system of the *Ārya Siddhanta*, it is..........=7471.993
The greatest precession of the equinoxes east and west, by the former, ... =24°
By the latter, it is also .... =24
The annual precession of the equinoxes, by the *Pārāsara Siddhanta*, ... =46''.5367
By the *Ārya Siddhanta*, it is .... =46.252
The year of the *Kali Yuga* when the precession was 0
by the former, was .... =3711, A.D. 610
by the latter, it was .... =3707, A.D. 606

From the very near agreement between the particulars above compared, no doubt whatever can exist, but that Āryabhatta was the real author of the *Pārāsara Siddhanta*, and for the express purpose above mentioned, that of supporting the imposition respecting the position and motions of the seven *Rishis*, invented by himself.

We shall now compute the position of the line of the *Rishis* by the system of Pārāsara. In 4320000000 years, there are 1599998 revolutions of the *Rishis* feigned; therefore each revolution $= \frac{4320000000}{1599998} = 2700$ years and a small fraction, which we reject.

At the commencement of the *Kalpa*, the line of the *Rishis* is assumed to be in the beginning of *Aswinī*; therefore, to find its position at the beginning of the *Kali Yuga*, we multiply the four right-hand figures of the revolutions 1599998, by 4567, reserving in the product the four right-hand digits, which will be the decimal parts of a revolution, showing the time then elapsed, since the line of the *Rishis* was in the beginning of *Aswinī*. Thus, $9998 \times 4567 = .0866$ of a revolution, the value of which in years $= 2700 \times .0866 = 233.82$, or near 234 years; that is, 234 years before the beginning of the *Kali*

Yuga, the line of the *Rishis* was in the beginning of *Aswinī*; consequently, at the beginning of the *Kali Yuga*, it was 34 years advanced into *Criticā*, from which point of time we compute as follows:—

Remainder of *Kriticā*,	. .	66 years.
Add for *Rohinī*,	. . .	100
for *Mṙigasiras*,	. .	100
for *Ardrā*,	. . .	100
for *Punarvasu*,	. .	100
for *Pushyā*,	. . .	100
for *Asleshā*,	. . .	100
Total to the beginning of *Maghā*,		666

That is to say, in the year 666 of *Kali Yuga*, or 2435 before the Christian era, the line of the *Rishis* was in the beginning of *Maghā*; which, therefore, as coming from the sage Pārāsara himself, who lived when the *Rishis* were in *Maghā*, was conclusive evidence of the truth of the time assigned to Pārāsara, Yudhisht'hira, &c. by Āryabhatta.

Having now explained the nature and object of Āryabhatta's imposition, and his forgery of the *Pārāsara Siddhanta* to support the same, we shall next proceed to show the age of the *Ārya Siddhanta* and *Pārāsara Siddhanta*, from the errors in the positions and motions of the planets, in the same manner as I have already done, in respect of the *Brahma Siddhanta* and *Surya Siddhanta*; for though we have the date of the *Ārya Siddhanta*, as given by its author, it may be so far satisfactory, to see it confirmed by computation.

We shall begin with the *Ārya Siddhanta*. The positions of the planets at sunrise, at the beginning of the *Kali Yuga*, on the meridian of *Ujein*, will be had by multiplying the three right-hand digits of each number of revolutions, as given in the table of the system, by 456, reserving the three right-hand figures in the product, which will be the planets' place in decimal parts of a revolution, thus: for Mercury, the number of revolutions in 4320000000 years=17937054671, the three right-hand figures 671, multiplied by 456, give .976, which reduced is $11^s 21^\circ 21' 36''$, the mean place of Mercury at the beginning of the *Kali Yuga:* or the same result will be obtained by saying, as 4320000000 : 17937054671, so the years elapsed of the *Kalpa* (in this case 1969920000), to the planets' mean place in revolutions, signs, &c.

The following were the mean places of the planets at the beginning of *Kali Yuga,* Hindu sphere, by Āryabhatta and La Lande's Tables.

Planets, &c.	By Aryabhatta.				By La Lande.				Differences.			
Sun, Hindu sphere,	0^s	0°	$0'$	$0''$	0^s	0°	$0'$	$0''$	0^s	0°	$0'$	$0''$
Moon,	0	0	0	0	0	3	8	48	0	3	8	48−
Mercury,	11	21	21	36	10	27	21	1	0	24	0	35+
Venus,	11	27	7	12	1	2	52	51	1	5	45	39−
Mars,	0	0	0	0	11	17	46	23	0	12	13	37+
Jupiter,	11	27	7	12	0	16	49	21	0	19	42	9−
Saturn,	0	0	0	0	11	8	46	40	0	21	13	20+
Moon's apogee,	4	3	50	24	3	29	58	19	0	3	52	5+
—— Node,	5	2	38	24	5	6	38	4	0	3	59	40−

The following are the mean annual motions of the planets, Hindu sphere, by Āryabhatta and La Lande's Tables.

	By Aryabhatta.				By La Lande.				Differences.	
Sun, ...	0^s	0°	$0'$	$0''$	0^s	0°	$0'$	$0''$	$0'$	$0''$
Moon, ...	4	12	46	40,5	4	12	46	37,7076	0	2,7924+
Mercury, ...	1	24	45	16,4013	1	24	45	36,1940	0	19,7927−
Venus, ...	7	15	11	51,4296	7	15	11	23,4870	0	27,9426+
Mars, ...	6	11	24	9,3	6	11	24	19,26	0	9,96 −
Jupiter, ...	1	0	21	5,9046	1	0	20	50,6965	0	15,2081+
Saturn, ...	0	12	12	50,7	0	12	13	9,5761	0	18,8761−
Moon's apogee,	1	10	40	32,6022	1	10	40	35,7990	0	3,1962−
—— Node, ..	0	19	21	34,0062	0	19	21	30,8420	0	3,1642+

Now, dividing the errors or differences in the mean places of the planets at the beginning of the *Kali Yuga*, by the errors or differences in the mean annual motions, we get as follow:—

Moon, ...	3°	8′	48″	} .....divided by 2″7924	= 4114 years.
Sec. equat. +		2	40		
Mercury, ...	24	0	35	.... by 19,7927	= 4367
Venus, ...	35	45	39	.... by 27,9426	= 4607
Mars, ...	12	13	37	.... by 9,9600	= 4419
Jupiter, ...	19	42	9	.... by 15,2081	= 4663
Saturn, ...	21	13	20	.... by 18,8761	= 4047
—— Apogee, ..	3	52	5	.... by 3,1962	= 4356
—— Node, ...	3	59	40	.... by 3,1642	= 4544
Their sum is					35117
And which, divided by 8, the mean is					4389

Differing only 34 years from the date 4423 given by Āryabhatta, and therefore a complete proof of the truth of it. Indeed there is no reason whatever to suppose that Āryabhatta, or any real Hindu astronomers, would falsify their own dates; nor indeed could they, because such an imposition would be observed by their contemporaries: the falsification of dates, and the interpolation of passages into books, with a view to give them or others the appearance of antiquity, are the artful contrivances of those that came after them. But though Āryabhatta, would not give a false date to his own work, yet he could forge a book in the name of another, as the *Pārāsara Siddhanta*, to serve his purpose.

The following Table, will now show the errors in the places of the planets by the *Ārya Siddhanta*, compared with La Lande's Tables, at different periods, from the beginning of the *Kali Yuga* to A.D. 1322.

Planets, &c.	Kali Yuga. or B.C. 3102.	Kali Yuga 1000.	Kali Yuga 2000.	Kali Yuga 3000.	Kali Yuga 4000.	K.Y. 4423, 1322 A.D.
Moon,	8°55′33″ –	6°10′15″ –	3°45′18″ –	1°45′ 9″ –	0°14′11″ –	0°14′22″ +
Moon's apogee	3 52 5 +	2 58 49 +	2 5 33 +	1 12 17 +	0 19 1 +	0 3 31 –
Moon's node	3 59 40 –	3 6 56 –	2 14 12 –	1 21 28 –	0 28 44 –	0 6 25 –
Mercury,	24 0 35 +	18 30 42 +	13 0 49 +	7 30 56 +	2 1 3 +	0 18 29 –
Venus,	35 45 39 –	27 59 56 –	20 14 13 –	12 28 30 –	4 42 47 –	1 25 47 –
Mars,	12 13 37 +	9 27 37 +	6 41 37 +	3 55 37 +	1 9 37 +	0 0 36 –
Jupiter,	19 42 9 –	15 29 9 –	11 16 9 –	7 3 9 –	2 50 9 –	1 2 56 –
Saturn,	21 13 20 +	15 58 44 +	10 44 8 +	5 29 32 +	0 14 56 +	1 58 8 –

Having thus proved the time the *Ārya Siddhanta* was written, independent of the date, we shall now proceed to show the age of the *Pārāsara Siddhanta* by the like process.

The positions of the planets at the beginning of the *Kali Yuga,* according to this system, will be obtained by the rule of proportion, in the usual manner, or more concisely, by multiplying the four right-hand figures of the number of revolutions for each of the planets, by 4567, reserving the four right-hand figures in the product, which will be decimal parts of a revolution, denoting the planets' mean place.

The following Table, shows the mean places of the planets at the beginning of *Kali Yuga,* Hindu sphere, at 6 A.M. on the meridian of *Ujein,* by the *Pārāsara Siddhanta* and La Lande's Tables, with the errors or differences.

Planets.	Pārāsara Siddhanta.				La Lande's Tables.				Errors or Differences.		
Sun, ...	0s	0°	0′	0″	0s	0°	0′	0″	0°	0′	0″
Moon, ...	0	0	10	48	0	3	8	48	2	58	0 –
Mercury, ...	11	21	17	16,8	10	27	21	0,9	23	56	15,9 +
Venus, ...	11	26	58	33,6	1	2	52	50,9	35	54	17,3 –
Mars, ...	11	29	14	38,4	11	17	47	22,9	11	27	15,5 +
Jupiter, ...	11	27	2	52,8	0	16	49	20,7	19	46	27,9 –
Saturn, ...	11	28	57	21,6	11	8	46	40	20	10	41,6 +
Moon's apogee,	4	5	12	28,8	3	29	58	19,2	5	14	9,6 +
——— Node,	5	2	49	12	5	6	38	3,8	3	48	51,8 –

The following Table, shows the mean annual motions of the planets in the Hindu sphere, according to the *Pārāsara Siddhanta* and La Lande's Tables, with the errors or differences.

Planets.	Pārāsara Siddhanta.	La Lande's Tables.	Errors or Differences.
Sun, ...	0ˢ 0° 0′ 0″	0ˢ 0° 0′ 0″	0″
Moon,........	4 12 46 40,3545	4 12 46 37,6236	2,7309 +
Mercury,......	1 24 45 16,6422	1 24 45 35,8979	19,2557 –
Venus, ...	7 15 11 51,6444	7 15 11 23,1328	28,5116 +
Mars, ...	6 11 24 9,9111	6 11 24 18,8799	8,9688 –
Jupiter, ...	1 0 21 5,9862	1 0 20 50,3068	15,6794 +
Saturn, ...	0 12 12 51,5439	0 12 13 9,1854	17,6415 –
Moon's apogee,	1 10 40 31,3902	1 10 40 35,4105	4,0203 –
—— Node, ...	0 19 21 33,9705	0 19 21 31,2355	2,7350 +

Having now the errors or differences in the mean places of the planets at the beginning of the *Kali Yuga,* and the errors or differences in the mean annual motions, let the former be divided by the latter, and we have,

Planets, &c.	Divisors.	Dividends.	Quotients.
Moon, ...	2″,7309	2° 58′ 0″	3910
—— Apogee, ..	4,0203	5 14 9,6	4688
—— Node, ...	2,7350	3 48 51,8	5020
Mercury, ...	19,2557	23 56 15,9	4475
Venus, ...	28,5116	35 54 17,3	4533
Mars, ...	8,9688	11 27 15,5	4597
Jupiter, ...	15,6794	19 46 27,9	4540
Saturn, ...	17,6415	20 10 41,6	4117
The sum of the quotients or results		..	35880

Which divided by 8, gives for a mean result 4485, or A. D. 1384; which, therefore, coupled with the comparisons already made, shows that it was the work of Āryabhatta, though it comes out a few years later than his date; but that can be of very little consideration here, for the object of the work was not extraordinary accuracy, but as an authority to support by *name* an imposition: indeed the nearer the system of the *Pārāsara Siddhanta* agreed with his own work, the more it became liable to suspicion; and I am therefore rather surprised that he did not use more artifice than to frame it for his own time.

The following Table, will now show the errors in the mean places of the planets by the *Pārāsara Siddhanta,* compared with La Lande's Tables at different periods, from the beginning of the *Kali Yuga* to A.D. 1322.

Planets, &c.	Kali Yuga, or B.C. 3102.	Kali Yuga 1000.	Kali Yuga 2000.	Kali Yuga 3000.	Kali Yuga 4000.	K. Y. 4423, 1322 A. D.
Moon,	8°44′45″ –	6° 0′28″ –	3°36′33″ –	1°37′25″ –	0° 7′29″ +	0°20′44″ +
Moon's apogee	5 14 10 +	4 7 10 +	3 0 10 +	1 53 9 +	0 46 9 +	0 17 49 +
Moon's Node	3 48 52 –	3 2 17 –	2 16 42 –	1 31 7 –	0 45 32 –	0 26 15 –
Mercury,	23 56 16 +	18 35 20 +	13 14 24 +	7 53 28 +	2 32 32 +	0 16 47 +
Venus,	35 54 17 –	27 59 5 –	20 3 53 –	12 8 41 –	4 13 29 –	0 52 29 –
Mars,	11 27 16 +	8 57 47 +	6 28 18 +	3 58 49 +	1 29 20 +	0 26 6 +
Jupiter,	19 46 28 –	15 25 9 –	11 3 50 –	6 42 31 –	2 21 12 –	0 30 40 –
Saturn,	20 40 42 +	5 16 40 +	10 22 38 +	5 28 36 +	0 34 34 –	1 29 48 –

As this Table will, I think, be sufficient to show, by inspection, the time nearly when the *Pārāsara Siddhanta* was framed, it must be altogether unnecessary to dwell longer on this point. We shall therefore now proceed to such other matters, contained in the *Ārya Siddhanta,* as may be deemed curious or deserving of notice.

In the twelfth chapter of the *Ārya Siddhanta,* Āryabhatta gives us a table of the longitudes and latitudes of the stars in the Lunar Asterisms. The longitudes, as usual, are all reckoned from the beginning of the Lunar Asterism *Aswinī,* as the commencement of the modern Hindu sphere, but determined by the points of the ecliptic, being cut by the circles of latitude, and not by the circles of declination; therefore, differing from the *Brahmā Siddhanta,* and others that follow the latter method. The author of the *Sarva Bhauma,* a modern writer, appears to have followed

Āryabhatta in his method: it will therefore be useful to exhibit both their tables at one view.

TABLE of the Longitudes and Latitudes of the Stars in Lunar Mansions, reckoned from the beginning of Aswini, *according to the* Arya Siddhanta *and* Sarva Bhauma Siddhanta.

Lunar Asterisms.	Ārya Siddhanta.		Sarva Bhauma.		Stars supposed to be meant.
	Longitudes	Latitudes	Longitudes	Latitudes	
1 *Aswinī*,	12° 0′	10° 0′ N.	12° 40′	10° 50′ N.	γ or β Arietis,
2 *Bharanī*,	24 30	12	25 8	12 55	35 Arietis,
3 *Křitikā*,	38 33	5	39 2	4 44	The Pleiades.
4 *Rohinī*,	47 33	5 S.	48 9	4 40 S.	Aldebaran.
5 *Mřigasiras*,	61 3	10	61 1	10 12	37, 39, 40 Orionis?
6 *Ārdrā*,	68 23	11	65 8	11 7	137 Tauri?
7 *Punarvasu*,	92 53	6 N.	94 53	6 0 N.	78 or 83 Geminorum.
8 *Pushyā*,	106 0	0	106 0	0 0	Præsepe, or γ, δ Cancri?
9 *As'leshā*,	111 0	7 S.	109 0	7 4 S.	49 or 60 Cancri.
10 *Maghā*,	129 0	0	129 0	0 0	Regulus.
11 *P. Phalgunī*,	140 23	12 N.	142 48	12 42 N.	70 or 71 Leonis.
12 *U. Phalgunī*,	150 23	13	156 0	13 55	β Leonis.
13 *Hastā*,	174 3	10 S.	175 13	12 0 S.	δ Corvi.
14 *Chitrā*,	182 53	2	183 50	1 52	Spica Virginis,
15 *Swāti*,	194 0	37 N.	182 24	41 5 N.	Arcturus?
16 *Visākhā*,	212 23	1 30 S.	212 36	1 25 S.	24 Libræ,
17 *Anurādhā*,	224 53	3	224 38	1 50	7 Scorpii.
18 *Jyesht'hā*,	230 3	4	230 5	3 28	Antares.
19 *Mulā*,	242 24	9	248 36	8 40	34 Scorpii?
20 *P. Āshād'hā*,	254 33	5 20	254 34	5 22	δ Sagittarii.
21 *U. Āshād'hā*,	260 33	5 S.	260 21	5 0 S.	φ Sagittarii.
Abhijit,	263 0	63 N.	262 10	62 14 N.	α Lyræ.
22 *Sravanā*,	280 3	30	280 3	30 5	γ or α Aquilæ.
23 *Dhanisht'hā*,	296 33	36	294 12	26 25	β or α Delphini?
24 *Satabhishā*,	319 53	0 20 S.	319 15	0 45 S.	λ Aquarii.
25 *P. Bhādrapadā*,	334 53	24	335 8	26 3	α Pegasi?
26 *U. Bhādrapadā*,	347 0	26	348 44	28 0	78 or γ Pegasi.
27 *Revatī*,	360 0	0	359 50	0 0	ζ Piscium.

On comparing the above latitudes and longitudes, some differences will be observed; but whether such differences arise from error, or from different stars being intended, cannot be so readily ascertained. Thus Āryabhatta gives the longitude of the star in the sixth mansion *Ārdrā*, 68° 23′; but in the *Sarva Bhauma*, it is set down 65° 8′, making a difference of upwards of 3°. The position of this star by the *Brahma Siddhanta* is 67° and 11″ S. from which data I find the longitude to be 65° 5′, and latitude 10° 50′ S. Hence it appears that the error is on the side of Āryabhatta, unless he meant another star; but this we can hardly suppose, because there is no other star 3° more to the east with the same latitude. On the other hand, Āryabhatta makes the longitude of the star in the seventh Lunar Mansion *Punarvasu*, 92° 53′. I make it, from the data in the *Brahma Siddhanta*, 92° 52′, differing but one minute; but the *Sarva Bhauma* makes it 94° 53′, differing 2°, which shows that there are errors on both sides, which, after all, may have arisen from carelessness in copying, and therefore it would be needless to offer any further observations on the differences in the table, as copies of the *Ārya Siddhanta* are not now procurable.

In the fifteenth chapter, Āryabhatta treats of the several rules of arithmetic, as addition, subtraction, multiplication, division, squares and cubes of numbers, and their roots, progressions, and other matters relating to the doctrine of numbers, after which he treats on geometrical problems; the whole of which appears to be the same, or very nearly so, and in the same order, as given in the *Lilāvati* of Bhāskara Āchārya; with this difference, that Āryabhatta only gives the problems and rules of solution generally, and without numbers, whereas in the *Lilāvati*, they are exemplified by figures and numercial solutions, so that the latter may be taken as a perpetual commentary on this part of the *Ārya Siddhanta:* but in the problems relating to the circle, of which there are about eighteen in the *Ārya Siddhanta*, Bhāskara Āchārya differs from him, particularly in the proportions of the diameter to the circumference. For instance, the diameter being given to find the circumference, one of Āryabhatta's rules is, "Multiply the square of the diameter by 10, the square root of the product is the circumference." Thus, suppose the diameter 10, then $10^2 \times 10 = 1000$, and $\sqrt{1000} = 31.6228$: this appears to be from Brahma Gupta. Bhāskara's rule is, Multiply the diameter by 3927, and divide by 1250, the quotient is the near circumference. Thus, suppose the diameter 10, then $\frac{39270}{1250} = \frac{3927}{125} = 31.416$. The following rule is given by Āryabhatta for finding the arc from the chord and arrow (or verse sine): Multiply the square of the arrow (v. s.) by 6, add the square of the chord, the root is the arc. Thus, suppose the diameter of a circle is 10, and we wanted to know the circumference, then 10 will be the chord and 5 the arrow; and we shall have $\sqrt{5^2 \times 6 + 10^2} = \sqrt{250} = 15.8114$ for the arc or semicircumference; its double, therefore $=$ 31.6228, the same as above from Brahma Gupta. The following is another of Āryabhatta's rules for finding the arc: Multiply the square of the arrow (v. s.) by 288, divide the product by 49, to the quotient add the square of the chord, the root of the sum is the near arc. Thus, suppose the chord $= 10 =$ the diameter of a circle, then the arrow will be 5, or the semidiameter, and $\frac{5^2 \times 288}{49} = 146.938$ and $\sqrt{146{,}938 + 10^2} = \sqrt{246.938} = 15.7143$, the arc of a semicircle, and its double 31.4286, the whole circumference of a circle whose diameter is 10; whence, the proportion is as 1 : 31.4286. But Āryabhatta, in the 17th chapter, in speaking of the orbits of the planets, gives us a nearer approach to the truth; for he there states the proportion as 191 to 600, or as 1 : 31.4136, which gives the circumference a small matter less than the proportion of Bhāskara in the *Lilāvati*. This, however, is not the invention of

Āryabhatta; for it is employed in the *Brahma Siddhanta, Surya Siddhanta,* and by all the astronomers before the time of Āryabhatta, as well as since, for computing the tables of sines, &c. though not immediately apparent. Thus, in computing the sines, they take the radius at 3438, and the circumference they divide into 21600′, the diameter is therefore 6876, hence the proportion is 6876 : 21600 : reduce these numbers to their least terms by dividing them by 36, the result will be 191 : 600, as stated by Āryabhatta.

Āryabhatta next proceeds to the doctrine of excavations, and the contents of solids; but unfortunately, after proceeding as far as *Chiti* (piles or stacks), the remainder of the chapter is lost, together with the whole of the sixteenth, which contained his algebra, and a few stanzas of the seventeenth—a loss the more to be regretted, as I fear it cannot be restored. If we had this part, we should have been able to ascertain what improvements were introduced by Bhāskara Acharya, who, it appears, lived 200 years after Āryabhatta, as will be shown in the next section. Āryabhatta, like many other Hindu writers, is now thrown back into antiquity, the cause of which will be explained when we come to speak of Bhāskara and Varāha Mihira in the next section.

SECTION IV.

Varāha Mihira, *like* Āryabhatta, *endeavours to support the new order of things — Perverts the meaning of a passage relating to the epoch of* Yudhist'hira, *who he places* 2448 *before Christ* — Varāha Mihira *mentions the Surya Siddhanta and* Āryabhatta — *States the heliacal rising of Canopus at Ujein, when the Sun was* 7° *short of Virgo — Gives the positions of the aphelia of the Planets in the Jātakārnava for the year* 1450 *Saca, or A. D.* 1528 — *The heliacal rising of Canopus at Ujein computed for that year, the result agrees with that which* Varāha *stated, being* 7° *short of Virgo — The point of heliacal rising of Canopus at Ujein, in the Hindu sphere, shown.*

THE next person we have to notice, in point of antiquity and celebrity, is Varāha Mihira. Like Āryabhatta, he contributed his mite towards supporting the modern order of things, by endeavouring to pervert the meaning of a passage respecting the epoch of Yudhisht'hira. When the modern system was first introduced, the epochs of ancient kings were referred to it, as they really stood, and amongst the rest, Yudhisht'hira, who lived about 575 years before Christ. The epoch or time of Yudhisthira ascending his throne, transferred to the modern *Kali Yuga*, corresponded to the year 2526. The meaning of this, Varāha Mihira thought fit to pervert, by saying that it was the number of years Yudhisthira lived before the era of *Saca;* placing him there for 2448 B.C. The era of *Saca*, it seems, then, fell very conveniently in his way; but if it had not, he would have hit on some other contrivance, equally as plausible and convincing. He also supported the notions of the motions of the stars in the Great Bear first broached by Āryabhatta, that of their being 100 years in each Lunar Asterism, and that they were in *Maghā* in the time of Yudhishthira. To dwell longer on such childish absurdities would be a waste of time: we shall, therefore, proceed to what is of much more importance, that of determining the period when Varāha Mihira lived, which will enable us to point out a great number of forgeries and impositions of the moderns, that were not in the least suspected.

I have noticed on a former occasion, that Varāha Mihira mentions the *Surya Siddhanta,* which would of course place him after the period to which that work has been referred. Varāha Mihira also mentions Āryabhatta; consequently he must have been posterior to him, that is, since A.D. 1322. But what he states himself is still better than all this, because it brings us at once to the point. He tells us in one of his works, the *Varāha Sanhita,* that Canopus rose *heliacally* at *Ujein,* when the sun was 7° short of Virgo; that is, when he was in 23° of Leo. This is a most important fact, because it serves to decide a point of time that has been long disputed.

In calculating the time, we may make choice of any particular year since the time of Āryabhatta: but as Varāha Mihira has given us his time in another of his works called the *Jātakārnava,* in which the positions of the aphelia of the planets for the year of *Saca* 1450, or A.D. 1528, are given, we may as well make choice of that year, because, if the *Jātakārnava* was the real work of Varāha Mihira, then we would naturally expect they would agree, or at least nearly so, and thereby they would confirm each other; but if they disagreed considerably in point of time, though passing under the name of the same author, we then certainly would have a right to conclude they were not written by one and the same person.

The longitude of *Canopus* in A.D. 1750 was..........	3ˢ	11°	31
Deduct precession for 222 years (=1750—1528,)= ..	0	3	3
Longitude of *Canopus* in A.D. 1528, =.............	3	8	27
Latitude of *Canopus*, = ...	0	75	51
Obliquity of the ecliptic,.....	0	23	30

From which we get the right ascension and declination of *Canopus* in A.D. 1528 by the following proportions:

1	As radius sine	90°	0′	10.
	Is to sine of the longitude of *Canopus*	81	33	9.9952597
	So cot. latitude of ditto	75	51	9.4015910
	To cot. of an ∠	76	0	9.3968507
	Subtract obliquity of the ecliptic	23	30	
	Remain	52	30	
2	As radius sine	90	0	10.
	Is to cos. longitude of *Canopus*, ...	81	33	9.1671586
	So cos. latitude.........................	75	51	9.3882101
	To cosine of...........................	87	56	8.5553687
3	As radius sine	90	0	10.
	Is to cos.	52	30	9.7844471
	So tan.	87	56	11.4426638
	To tan.	86	36	11.2271109
	Which taken from	180	0	
	Leaves the right ascension of *Canopus*	93	24	in A.D. 1528.
4	As radius sine	90	0	10.
	Is to sine	52	30	9.8994667
	So sine	87	56	9.9997174
	To sine of the declination of *Canopus* = ..	52	27	9.8991841

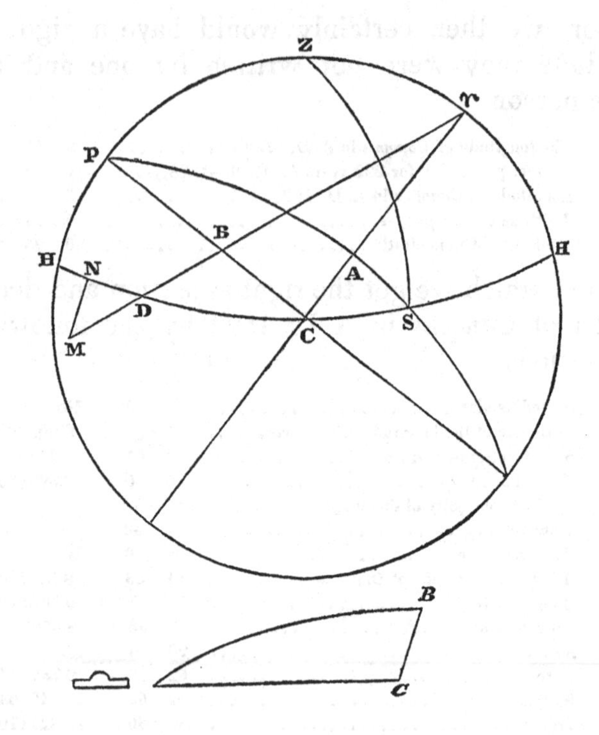

Now, in the annexed Diagram,

Let P be the pole of the equator.

PZ = 90° − 23° 11′ = 66° 49′ = the complement of the Lat. of *Ujein.*

♈ C the equator.

♈ D the ecliptic.

S place of the star.

Then say:

5	As tan. co. lat. PZ = ACS	66°	49′	10.3682963
	Is to rad. sine	90	0	10.
	So tan. of the declination AS	52	27	10.1142350
	To sine of the ascensional difference AC	33	51	9.7459387
	Add right ascension	93	24	
	The sum is the oblique ascension ♈ C	127	15	
	Which taken from	180	0	
	Leaves C ♎ (supplementary triangle,)	52	45	
6	As cos. of the obliquity of the ecliptic ∠ ♎	23	30	9.9623978
	Is to rad. sine	90	0	10.
	So tan. C ♎	52	45	10.1189478
	To tan. B ♎	55	7	10.1565500

And 180°—55° 7′ = 124° 53′ = ♈ B, the point of the ecliptic B, or the longitude of that point of the equator C, which rises with Canopus.

Now the next portion of the ecliptic to be found is BD.

7	As radius sine	90°	0′	10.
	Is to sine C♎	52	45	9.9009142
	So tan. obliquity of the ecliptic ∠♎	23	30	9.6383019
	To tan. BC	19	6	9.5392161
8	As rad. sine	90	0	10.
	To sine of the obliquity of the ecliptic ∠♎	23	30	9.6006997
	So cos. C♎	52	45	9.7819664
	To cos. CB♎	76	2	9.3826661
9	As rad. sine	90	0	10.
	Is to cos. BC	19	6	9.9754083
	So tan. of the latitude of *Ujein*	23	11	9.6317037
	To cot.	67	58	9.6071120
	Which taken from 76° 2′ leaves	8	4	
10	As cos.	8	4	9.9956815
	Is to cos.	67	58	9.5742003
	So tan. CB	19	6	9.5394287
	To tan. BD	7	29	9.1179475

To which add ♈ B = 124° 53′ we have the longitude of the point D = 132° 22′ coascendent with S and C, or the cosmical point of rising of Canopus.

The next portion of the ecliptic to be found is DM, depressed 10° 30′ below the horizon.

11	As rad. sine	90°	0′	10.
	Is to cos. BD	7	29	9.9962852
	So tan.	8	4	9.1514543
	To cos. CDB	82	0	9.1477395
12	As sine CDB = NDM	82	0	9.9957528
	Is to sine MN	10	30	9.2606330
	So is rad. sine	90	0	10.
	To sine DM	10	36	9.2648802
	Add the point of cosmical rising	132	22	
	The sum is the heliacal point	142	58 = 4ˢ 22° 58′, or 22° 58′ of Leo,	

which is within two minutes of what Varāha states it, he making it 23° of Leo, or 7° short of Virgo.

In the above calculation, I have taken 10° 30′ for the arc of vision, or the depression of the sun below the horizon at the rising of *Canopus*. Some allow 10°, others 11°; mine is a medium between the two, and agrees best with what Varāha states: but whether we take 10° or 11°, the difference will be inconsiderable, and not exceeding 30′ in the sun's longitude either way. Therefore the question respecting the true time of Varāha Mihira is now finally settled, and all doubts respecting his being the author of the *Jātakārnava* proved to be entirely groundless.

It may not now be amiss, to show the points in the Hindu sphere, to which the sun must have come at the cosmical and heliacal rising of *Canopus* at *Ujein* in A. D. 1528, as it will serve to prevent misconception. This may be done by deducting the difference between the Hindu and tropical spheres for the year A.D. 1528, from the cosmical and heliacal points 4^s 12° 22′ and 4^s 22° 58′ above determined. But as this would require some calculation, I shall show how it may be done without any trouble, by means of a celestial globe, and that independent altogether of any reference to the tropical sphere.

Take a celestial globe, on which the stars are correctly laid down, and rectify it for the latitude of *Ujein* 23° 11′ N. bring the star *Canopus* to the eastern horizon, and mark the point on the ecliptic then in the horizon, with its distance from some fixed star east or west of it, lying in or near the ecliptic; this will be the cosmical point, and its longitude in the Hindu sphere, will be known from its distance in degrees east or west of the star. Measure 10½° towards the east, perpendicular to the horizon, and mark the point where it falls on the ecliptic, in respect to its distance east or west of some fixed star whose longitude is given in the Hindu sphere, this will be the heliacal point, and its longitude in the Hindu sphere will be known from its distance east or west of the star. Thus, the globe being rectified for the latitude, and the star *Canopus* brought to the eastern horizon, the point of the ecliptic then on the horizon will be found to be about 10° 56′ west of Regulus, whose longitude in the Hindu sphere is 4^s 9°; consequently the longitude of the cosmical point in the Hindu sphere, is $4^s\ 9° - 10°\ 56' = 3^s\ 28°\ 4'$: now measuring 10½° perpendicular to the horizon towards the east, as directed, the point will fall on the ecliptic about 20′ to the west of Regulus; therefore the longitude of this point in the Hindu sphere, is $4^s\ 9' - 0°\ 20' = 4^s\ 8°\ 40'$, to which the sun must invariably come at the heliacal rising of *Canopus* at *Ujein*. The Hindu sphere being sideral, and consequently fixed, the cosmical and heliacal points thus shown, are also nearly fixed: they were so in the time of Varāha, and they are the same now. But it is far otherwise in the moveable or tropical sphere, in which the variation is considerable, the longitude of the cosmical and heliacal points increasing as the time is more modern.

Having thus shown the age in which Varāha Mihira wrote, from the data given in both his works, the *Varāhi Sanhita* and *Jātakārnava*, which completely agree, we shall, in the next section, endeavour to explain the cause of his being thrown back into antiquity by the moderns, with the various means that have been employed for that purpose.

SECTION V.

The cause of Varāha Mihira *being thrown back into antiquity by the moderns explained—The reason of two* Varāha Mihiras *and two* Bhāskaras *explained by the imposition on* Akber—Bhāskara *thrown back to A.D.* 1150—*A number of forgeries to support the imposition—Spurious Ārya Siddhanta—Two Bhāsvatis—Pretended ancient commentaries—Interpolations—The Pancha Siddhantika—False positions of the Colures—Artificial rules for the cosmical rising of Canopus by the Bhāvasti—By the Pancha Siddhantiki—By* Kesava—*By the Graha Lāghava—The time to which they refer appears to be about the middle of last century—The heliacal rising of Canopus by the Brahma Vaivarta, and Bhavisya Puranas, when the sun was* 3° *short of Virgo—A view of the impositions arising from spurious books*—Laksmidasa, *a commentator on the Siddhanta Siromani, pretended to be a grandson of* Kesava, *and to have written in A.D.* 1500—*Determines the cosmical rising of Canopus at Benares for that year—The spurious Ārya Siddhanta examined, and shown to be a modern forgery—The system it contains, how framed—Gives the proportion of the diameter of a circle to its circumference the same as* Bhāskara—*Quotes the Brahma Siddhanta,* Brahma Gupta, *and the Surya Siddhanta—The Pulisa Siddhanta, another forgery, noticed—Forgeries of books innumerable—The Brahma Siddhanta Sphuta, another forgery—The object of the forgery—The spurious Brahma Siddhanta quotes the spurious Ārya Siddhanta, Pulisa Siddhanta, and Varāha Mihira, thereby proving it to be a forgery, and, perhaps, by the same individual—Quotations made from it to show the same—Mistake about the positions of the Colures, and the meaning of the terms Aries, Taurus, &c.—Interpolations—Vishnu Chandra, &c.*

About the year A.D. 1556, the emperor Jelaledeen Mahomed Akber ascended the throne. This prince was universally esteemed as a great encourager of learning, and of learned men of all nations; in consequence of which, a number of works on various subjects were continually presented to him, and among these was the *Lilāvati* of Bhāskara Āchārya. This work, in order to increase its value, as well as to exalt the abilities of the Hindus as men of science in the eyes of the emperor, was given out to be then several centuries old; and that it was not the work of Bhāskara ĀCharya, who was then living, but of another person of the same name, who lived as far back as the year A.D. 1150. By this Brahminical contrivance, two Bhāskara Āchāryas were framed out of one. But it so happened that Bhāskara Āchārya, in one of his works, the *Siddhanta Siromani,* mentioned the name of Varāha, who appears above to have written only twenty-six years before Akber ascended the throne, and consequently might then be still living. This untoward circumstance was, therefore, likely to overturn the whole imposition; and, if observed, instead of getting favour with the emperor, they would be considered as cheats and impostors: therefore, to save appearances and their credit, the same method was followed as with Bhāskara, by giving out that the person mentioned by Bhāskara was not the Varāha who wrote a few years before the emperor ascended the throne, but another of the same name, who lived about the time of Raja Bhoj, or his successor, by which means the discovery of the deception was prevented, and two Varāha Mihiras were thus made out of one.

It would appear that the matters remained in this state, depending entirely on mere verbal assertions, until the last century, perhaps about the middle of it, when the impositions above stated, were most probably opposed, and nearly overturned. Who the persons were that made the opposition we cannot now ascertain, nor the proofs produced, except by inferences from the various means that appear to have been employed to counteract such proofs. The time of Āryabhatta was known from the date of the *Ārya Siddhanta,* which was still existing; and Bhāskara Āchārya had written a commentary on the *Ārya Siddhanta,* which is supposed to have been the foundation of the *Lilāvati:* consequently Bhāskara must have been posterior to Āryabhatta. Āryabhatta wrote the *Ārya Siddhanta* in the year A.D. 1322, (Sec. III.) but Bhāskara Āchārya, (in order to make the *Lilāvati* appear ancient,) was thrown back to the year 1150, or 172 years before the very person on whose work he wrote a commentary. Here was a most glaring inconsistency, that could not be otherwise than noticed, and as such, fatal to the story of the pretended antiquity of Bhāskara Ācharya, as well as to that of Varāha Mihira. Moreover, the positions of the planets given by the *Ārya Siddhanta* proved the truth of the date of that work, so that there was no room left for subterfuge. These being the proofs, I suppose, that were brought forward, or at least a part of them, we shall now endeavour to show the means that were adopted to counteract them. It was evident, that so long as the *Ārya Siddhanta* was in existence and in circulation, any means they might adopt to counteract the proofs would be useless. It became, therefore, necessary by all means to suppress it, and to fabricate a variety of books, and make interpolations in existing ones, for the express purpose of supporting the antiquity they had thus given, or meant to give, to Varāha Mihira and Bhāskara Āchārya. Among the number of books that were thus fabricated to answer this purpose, we accordingly find a spurious *Ārya Siddhanta*, as a substitute for the real one suppressed, but differing entirely from it and appearing an imposition on the face of it. We also find a book called tbe *Kutuhala,* fathered on Bhāskara Āchārya, in which the epoch for the positions of the planets is given for the year 1105 *Saca,* by which it was intended that the world should believe that he actually lived at that period, and wrote the work. Other works were written to throw back the time of Varāha Mihira, such as the two *Bhāwatis*, pretended to have been written by Satanunda, the fictitious pupil of Varāha; the epoch of these is the year 1021 *Saca*. One is made to mention the name of Varāha, and the other to allude to the five *Siddhantas*, which it was pretended Varāha had written, or compared and examined: consequently, from the epoch thus given, it was intended to establish that Varāha Mihira lived as far back, at least, as the year 1021 *Saca*, or A.D. 1099. Besides these, there is a commentary on the *Varāhi Sanhita*, by some person or persons, under the fictitious name of Bhattotpala, given out by the Brahmins of *Ujein* to be near eight hundred years old; for this Bhattotpala is pretended to have lived in A. D. 968*: consequently the assertions of the Brahmins of *Ujein*, if believed, would place Varāha Mihira anterior to A.D. 968; for no one will doubt, but that he must be older than his commentator. And to strengthen all this, they had recourse to several interpolations, which they made into the original work of Varāha Mihira, the *Varāhi Sanhita*, one of which, in particular, we shall here notice, to show the artfulness of the method that was adopted.

They feigned that Varāha Mihira had examined and compared five certain *Siddhantas*, or works on astronomy, called by the names of the *Brahma*, the *Surya* the *Vasishtha*, the *Romaka*, and the *Pulisa Siddhanta*, which names they interpolated into the *Varāhi Sanhita*, to support the fiction. They also pretended that Varāha Mihira wrote a work on these

* Mr. Colebrooke's Notes and Illustrations, p. xxxiii.

five *Siddhantas*, entitled the *Pancha Siddhantika*, but which work it does not appear that any one has ever yet seen. From this unseen and unheard of work, and which I am disposed to think, never existed, Bhattotpala, the commentator on the *Varāhi Sanhita*, or rather the person who assumed that name, pretends to quote a passage, stating that the colure was in the Lunar Asterism *Punarvasu* when the author wrote that work; which, as no particular degree is mentioned, would place him between the years A.D. 277 and A.D. 1165. This forgery is, however, completely overthrown, as well as all the others we have mentioned, by the time deduced from the heliacal rising of the *Canopus* mentioned by the author himself in his *Sanhita*, and by the time for which he gave the positions of the aphelia of the planets in his other work, the *Jātakārnava*.* It

---

* Mr. Colebrooke is not disposed to believe the *Jātakārnava* a work of Varāha Mihira, because it forms a direct positive proof against him. He says: " The " minor works ascribed to the same author (Varāha Mihira), may have been com- " posed in later times, and the name of a celebrated author have been affixed to " them, according to a practice which is but too common in India, as in many other " countries. The *Jātakārnava*, for example, which has been attributed to him, may " not improbably be the work of a different author. At least I am not apprized " of any collateral evidence, such as quotations from it in books of some antiquity, " to support its genuineness as a work of Varāha Mihira."—As. Res. Vol. XII. p. 244. The genuineness of the *Jātakārnava* is proved by its agreement in date with the *Varāhi Sanhita*, to both of which the author's name is affixed. Quotations can never be considered as proof, for they may be, and often are made by impostors to give strength to their forgeries; and what is more, spurious works very seldom quote any other but those of the same character, lest a discovery of the real truth should be made by that means. For instance, the spurious *Ārya Siddhanta* is repeatedly quoted by Bhattotpala, (one of the greatest impostors that ever lived), and also by the spurious *Brahma Siddhanta*, but not a word about the genuine *Ārya Siddhanta*, which, however, is found quoted in other books. And when quotations are made by impostors from genuine books, it is generally with a view of introducing to notice some passage that has been interpolated for the purpose of imposition. Thus the *Pancha Siddhantika* is quoted by Bhattotpala, to bring forward the pretended positions of the colures in the time of Varāha Mihira, with the intention of throwing that author back into antiquity; which, however, is overthrown by the other passage he cites from the same book relative to the cosmical rising of *Canopus*,

might, therefore, be considered superfluous to enter into all these particulars, since every circumstance

---

showing it to be the production of the last century; and consequently all the interpolations into the *Varāhi Sanhita* relating to it, to *Pulisa*, and the spurious *Ārya Siddhanta*, are of the same period. But it is upon these quotations, interpolations, and assertions without proof, that Mr. Colebrooke relies, and not on astronomical facts, which he seems to disregard. He tells us, that Varaha Mihira was the son of Adityadasa, being so described in the *Vrihat Jātaka*, of which he was the author, and that he lived in the fifth or sixth century of the Christian era. Notes and Illustrations, pp. xlv. li.—That Adityadasa might possibly be the father of Varaha Mihira, for any thing we can know to the contrary; but that he flourished at the close of the fifth, or beginning of the sixth century of the Christian era, is what cannot be admitted, because we have already shown him, from astronomical facts, to have been contemporary with the emperor Akber. Whether he was the author of the *Vrihat Jātaka* or not, cannot possibly be known, except from circumstances; for Mr. Colebrooke himself has admitted the practice of India and other countries, of naming books on celebrated authors that never saw them. And the assertion of Bhattotpala, the impostor, making him the author of it, is here of no weight. Mr. Colebrooke also states, that "Varaha Mihira is cited by name in the *Pancha Tantra*, the original of the Fables of Pilpay, which were translated into Persian for Nushervan, more than 1200 years ago." As. Res. Vol. IX. p. 364.—All this proves nothing. For it does not follow that the *Pancha Tantra* of Vishnu Sarmana is the original of the Fables of Pilpay: on the contrary, it is more likely that the latter should be the original of the former, and perhaps of many others, under different names, in the same manner as we have a number of *Rāmāyanas*, and other books on subjects of the same kind. We have no proof whatever that the *Pancha Tantra* of Vishnu Sarmana is even a hundred years old: for, to prove that it was the identical one that was translated into Persian more than 1200 years ago, it ought be shown, that the name of Varaha Mihira, was actually in that very translation, and still continues, without which it could be no proof; for we know that all Hindu books are liable to interpolations, and consequently the *Pancha Tranta* as much so as any other. But, supposing the fact was true, that the name of Varaha Mihira was actually in the original translation into Persian upwards of 1200 years ago, what would that prove, after all? It could only go to prove that there have been more persons of the name of Varaha Mihira than one; but it would never affect the time of the author of the *Varāhi Sanhita*, who was the contemporary with Akber. Mr. Colebrooke himself, seems to say, there have been more than one Varaha Mihira, (Notes and Illustrations, p. li.) but I have not seen as yet any sufficient proof to that effect, and therefore I am disposed to believe there was but one, and to consider all the rest as arising from the effect of imposition. I have seen an old copy of the *Vrihat Jātaka* above alluded to by Mr. Colebrooke, and it does not contain a syllable of what he states, as to Varaha Mihira living in the fifth or sixth century of the Christian era, &c. I think Mr. Colebrooke, like my old friend, the late Colonel Wilford, and perhaps many others, was imposed on by his crafty dependants, who studied his inclinations and his wishes, and, from knowing the bias of his sentiments, were thereby enabled to practise, with security and advantage to themselves, their imposture of forged and interpolated books, which they produced for him, or put in his way to obtain, as might appear best to answer their purpose.

that does not accord with these decisive facts, must be taken and considered as an absolute forgery or imposition ; but it may be useful to point the artful manner by which they were conducted, which never could have been fully detected, had it not been for the circumstance of the heliacal rising of *Canopus*, given by Varaha Mihira himself, of the effect of which, perhaps, they were not fully aware, otherwise they would in all probability have expunged it: but so it always happens, that impostors, however careful they may be, in giving a plausible appearance to their writings, yet, overlook some circumstance, which entirely overthrows the whole fabric; as may be seen by the instance above given. But it is not the above instance alone that we are to notice; there are many others, some of which I shall now bring forward to support the same facts.

THE BHASVATI.

This work I have noticed above, as being pretended to have been written by a person of the name of Satananda in the year A. D. 1099, who is also said to have been a pupil of Varāha Mihira: therefore, by this contrivance, Varāha Mihira would be placed anterior to that period. But the author of this work, whatever his real name may have been, has left us his rule for computing the cosmical rising of *Canopus* in his own time, from which we are enabled to judge whether the date assigned be true or false. His rule is thus given: Multiply the length of the equinoctial shadow by 25, to the product add 900, (the sum will be degrees, of which 225 are equal to one sign); when the sun is in the longitude expressed by this quantity, then the star

Canopus rises cosmically, or at the same time with the sun.

Example:—The length of the equinoctial shadow at *Ujein*, cast by a gnomon of 12, is taken at . . 5
Multiply this by 25, the product is . . 125
Add 900, the sum will be 1025
Which divided by 225 for each sign, we get 4^s 16° 40′ for the sun's longitude at the time of the cosmical rising of *Canopus* at *Ujein*, which, by adding 10° 20′, would give the heliacal rising the same as in the *Brahma Vaivarta Purana, &c.*

THE PANCHA SIDDHANTIKA.

This is quoted by the commentator Bhattotpala, as one of the works of Varāha Mihira, and from which he cites a similar rule to the preceding for the cosmical rising of *Canopus*, which is thùs given: Multiply half the equinoctial shadow by the square of 5, add to the product 73, the sum will be degrees and minutes; and when the sun's longitude is equal to the sum, then *Canopus* rises cosmically, or at the same time with the sun.

Example:—Length of the equinoctial shadow at *Ujein* is 5, and its half= . . . . $2\frac{1}{2}$
Multiply by the square of 5=25, the product is . . . . 62° 30′
Add 73, and the sum is 135° 30′=4^s 15° 30′, the sun's longitude when *Canopus* rose cosmically at *Ujein*.

KESAVA.

Kesava's rule is precisely the same as the *Pancha Siddhantika*, and therefore gives the same longitude to the sun, 4^s 15° 30′, at the time of the cosmical rising of *Canopus*.

THE GRAHA LAGHAVA

Gives the following rule:—Multiply the equinoctial shadow by 8, to the product add 98, the sum will be the sun's longitude when the star *Canopus* rises cosmically, or in the morning with the sun: and subtracting the product from 78, shows the sun's longitude at the heliacal setting.

Example:—The equinoctial shadow at *Ujein*	5
Multiply by 8, the product is . .	40
* Add to the product	98
The sum is	138

Or 4^s 18° the sun's longitude when *Canopus* rose at sunrise, and 78°—40°=38°, the sun's longitude, when that star set heliacally at *Ujein*.

The artificial rules for the cosmical rising of *Canopus* by the different books being now explained, we shall next show the time to which they refer. In A.D. 1750, the cosmical rising of *Canopus* at *Ujein* was when the sun's longitude was 4^s 13° 34′, which is less than any of the Hindu books make it. The cause of this is owing to their method of estimating it from the sun's longitude, at the time of heliacal rising of *Canopus*, by allowing a deduction of about 10 degrees. Thus the sun's longitude at

* The degrees added in these operations do not arise from any scientific principles; they are only the differences to make up the sun's longitude to what it is required to be at the cosmical rising. Thus, suppose I know that the sun's longitude at the cosmical rising of *Canopus* is 4^s 17° at *Ujein*, then I may multiply the length of the equinoctial shadow by 5, 6, 8, 10, or any other number, at pleasure, and add the difference to make up 4^s 17°; for the rules are entirely artificial.

the time of heliacal rising of *Canopus* in A. D. 1750, was about $4^s\ 25°\ 40'$; deduct from this 10° or $10\frac{1}{2}°$, the remainder is $4^s\ 15°\ 10'$: the *Pancha Siddhantika* and *Kesava* make it $4^s\ 15°$; therefore, making every possible allowance for error, they cannot be older than the middle, or at farthest the beginning of last century. And the differences between the supposed dates of the books mentioned above, were doubtless made on purpose as a deception, to give, as it were, the appearance of different degrees of antiquity to the authors, while, in fact, they are all of the same period, or at least nearly so; the whole, with many others, being probably the contrivance of a junta of Brahmins during the last century.

Here, I think, we have a complete proof of the imposition of the *Brahmins* of *Ujein*, in telling the late Dr. Hunter that Bhattotpala, who cites the *Pancha Siddhantika*, lived near 800 years ago. Certainly *Ujein* is not the place to obtain any true information on matters of this kind, it being the very focus or point from which the whole of the impositions appear to have originally sprung; though at present, I believe, the fabrication of spurious books, commentaries, &c. is not confined to *Ujein*, but to be found all over India, wherever the influence of the *Brahmins* prevails.

THE BRAHMA VAIVARTA AND BHAVISYA PURĀNAS.

These books state the heliacal rising of *Canopus* when the sun is 3° short of Virgo, or $4^s\ 27°$ of Leo: they are therefore the productions of the same period; and thus we may see that the Purānas are not so ancient as generally supposed, or the Hindus and others would wish us to believe.

The epoch of the two *Bhāswatis* is the year 1021 *Saca*, or A.D. 1099. Bhattotpala, according to the information received by Dr. Hunter when at *Ujein*, was said to have lived about the year 890 of *Saca*, or A.D. 968; Kesava about 1382 *Saca*, or A.D. 1440; and the epoch of the *Grahalaghava*, feigned to be written by Ganesa, the son of Kesava, 1442 *Saca*, or A.D. 1520. And yet none of them, from what has been shown above, can be referred further back than the middle or beginning of last century. Hence we can depend on neither names, dates, nor epochs in Hindu books. One Gungādhur is said to have written a commentary on the work of Bhāskara Āchārya in the year 1342 *Saca*, or A.D. 1420, which we know, from the time of Varāha Mihira, A.D. 1528, already determined, to be impossible, in respect of date. Laksmidas, the feigned grandson of Kesava, wrote a commentary on the *Siddhanta Siromani* of Bhaskara, in which all the calculations of the places of the planets, and the cosmical and heliacal risings of two of the stars, are made for the end of the year 4601 of the *Kali Yuga*, or A.D. 1500. This, no doubt, was done with a view to make it believed that he lived and wrote at the very epoch for which he made his calculation: but how could this be, if his supposed grandfather, Kesava, lived only about the middle of last century, or say, even in the beginning of it, at farthest? In fact, names, dates, and astronomical epochs, are given to books, at the will and pleasure of the writers, and seldom have any reference to the names, dates, and epochs of the real authors.* The

---

* I mentioned this circumstance to a Hindu astronomer, who acknowledged the fact, and, in defence of it, said: " Some men make the commencement of the *Kalpa*,

object here, as well as in all the rest was to support the pretended antiquity of Varāha Mihira, and of Bhāskara Āchārya, the latter of whom the author states for that purpose, as having been born in the year 1036 of *Saca* (A.D. 1114), and of his having written the *Siddhanta Siromani* in the year 1072 of *Saca*, at the age of 36. The work, however, though intended to support an imposition, is useful in other respects: it gives the different operations for determining the mean and the true places of the planets, both heliocentric and geocentric; the calculations of lunar and solar eclipses; the rising and setting of the planets, &c. together with the cosmical and heliacal rising of *a* Delphini and *Canopus*, by which it is shown, that the sun's longitude at the times of the cosmical and heliacal rising of the stars is reckoned from the vernal equinoctial point, and not from the beginning of *Aswinī*, or the *Hindu* sphere. Thus, in calculating the cosmical and heliacal rising

---

the epoch from which they direct the calculations of the places of the planets to be made for any time required. Others make choice of the beginning of the *Kali Yuga* for the same purpose. Have we not, therefore, a right to make choice of any other epoch, such as 2, 3, 4, or 5 hundred years or more, back from our own times, for the like purpose?" To which I replied, that certainly they had; but that, in consequence of the epoch's being brought down to modern times, and within the scale of probability, others were deceived by it, and supposed the author must have lived at the epoch for which he gave the positions of the planets. To which he replied, that that was not their fault; for that there was no absolute connection whatever between the epochs and the times of the authors: that they might be the same or not, as they found it best to answer the purpose: that in astronomical books, the principal object was to give rules for determining the places of the planets for any time required, from the motions and positions given at some certain epoch, without any regard to the time of the author. I asked him, if he framed an astronomical work, and placed the epoch from which the calculation should proceed 1000 years back, if he would put his name to it, and own the work. He replied, he could do so, on the principle already explained; but, in such cases, it was usual to put the name of some ancient sage to it, or that of some fictitious astronomer, with an account of his birth, parentage, connections, and country, in order to give it the plausible appearance of being ancient and real, which, according to modern notions, would much enhance its value.

of *Canopus* for the 4602d year of the *Kali Yuga*, or A.D. 1500, he takes tabular longitude, or the distance between the beginning of *Aswinī* and the point of the ecliptic cut by the circle of declination of the star, which is . . . . . . 87°

He takes the tabular latitude, or distance between the same point and the star, counted on the circle of declination, which is 77

From these he determines, by approximations, the true longitude, latitude, and declination. That is to say, the longitude, reckoned from the beginning of *Aswinī*, to the point in the ecliptic cut by the circle of latitude, which he makes . . 80° 39′ 11″

Making a difference between the points of the ecliptic cut by the circle of declination and that of latitude . 6 20 49

The true longitude of the star from the beginning of *Aswinī* being thus determined, he then adds the precession of the equinoxes, down to the end of the year 4601 of the *Kali Yuga* (A.D. 1500) 16° 10′ 59″, the sum gives the longitude of the star in A.D. 1500, reckoned from the vernal equinoctial point, equal to . . . . 96° 50′ 10″

He then determines the sun's longitude at the cosmical and heliacal rising of the star at Benares, where the length of the equinoctial shadow is found to be 5 50″ = 5.75, or latitude north . 25° 36′ 8″

He makes the cosmical rising, when the sun's longitude was . . . . 4^s 16°

The heliacal rising at 4 25 36 9

All of which are reckoned from the vernal equinoctial point, or beginning of Aries, and not from the beginning of the Hindu sphere.

It is, however, to be observed, that there is an error in the computed longitude of *Canopus* used, which affects all the calculations connected with it. The longitude of *Canopus* in A. D. 1500, at the end of the year 4601 of the *Kali Yuga*, by

European Tables, was .	3^s	8°	1′	30″
But by the Hindu computation, it was only	3	6	50	10
Difference, the latter too little by	0	1	11	20

Which must, of course, affect the results of the calculations.

In A. D. 1500, the latitude of *Canopus* was	75°	51′	0″
The right ascension . . .	93	12	33
The declination	52	26	0
And the obliquity of the ecliptic .	23	30	22
If we now take the latitude of Benares at what the Hindus state it, from the length of the equinoctial shadow, as	25	36	8

The cosmical rising of *Canopus* in A. D. 1500, will come out . . .	4^s	17°	6′	34″
Differing from the Hindu computation by	0	1	6	34

THE SPURIOUS ĀRYA SIDDHANTA.

This trifling production, has been substituted in the place of the real one, now suppressed (see Sec. III.): but notwithstanding all the care that has been taken to prevent detection, appears to be a forgery of very modern date. In this, as well as in the works above mentioned, care had been taken that the positions of the planets at the epochs a sumed, should be made to agree with the times; for they discovered, from the circumstances of the *Pārāsara Siddhanta* (given in Sec. III.), that pretending a book to be ancient would not alone be sufficient, as it would be liable to be detected by the positions of the planets not agreeing with such pretended antiquity, but to a more modern period, the time when the work was actually written; and therefore, in computing the places of the planets for forged epochs of antiquity, they perceived the necessity of computing from some tables the actual positions for the time. It is not, however, on the mere position of the planets alone that we are to depend; for these may be so accurately determined for the time, by computation, as completely to baffle all our attempts in detecting the fraud. We are, therefore, to have recourse to other means, and to consider every circumstance that serves to point out the work to be more modern than the period to which imposition has referred it, as we have done with the books above mentioned, which, by means of the cosmical risings of *Canopus*, are decisively proved to be forgeries in respect to their dates, epochs, names, &c.

The revolutions of the planets given in the spurious *Ārya Siddhanta*, have no affinity whatever with those given in the real *Ārya Siddhanta:* they appear to have been computed from those given in the *Surya Siddhanta.* The revolutions of the planets given in the *Surya Siddhanta,* answered when that work was written; but in process of time, they were found to require corrections, to make them agree with the actual positions of the planets corresponding to the time; so that about 260 or 300 years ago, it was found necessary to make the following corrections, which the Hindu astronomers call *bij:*—

Mercury	required a correction of	16 revolutions in 4320000 years, subtractive from		17937060
Venus		12 ditto	....	7022376
Mars		0	....	2296832
Jupiter		8	....	364220
Saturn		12	.... additive to	146568
Sun		0	....	4320000
Moon		0	....	57753336
—— Apogee		4	.... subtractive from	488203
—— Node		4	.... additive to	232238

the revolutions on the right hand of the page, being those given in the *Surya Siddhanta,* for 4320000 years. The *Surya Siddhanta,* from what is shown in Section II. appears to have been written about the year A.D. 1091; therefore, in framing numbers to answer for a period anterior to A.D. 1091, we would apply the corrections with a contrary sign, thus: the number of revolutions of Venus in 4320000 by the *Surya Siddhanta* = 7022376; but this number in modern times (about A.D. 1550), required to be diminished by 12, to render the result correct: consequently, in going back from A.D. 1091, the number would require to be increased, instead of diminished; so that, if the distance of time was the same, the number would come out 7022376 + 12 = 7022388, which is the number given in the spurious *Ārya Siddhanta.* Other means, such as European or Mahomedan tables, might have been also used for the more effectual correcting the numbers; but this explanation is sufficient.

The following Table exhibits the numbers, according to the *Surya Siddhanta* and the spurious *Ārya Siddhanta,* with their differences.

	Surya Siddhanta.		Spurious Ārya Siddhanta.	
Sun,	4320000	.	4320000	difference 0
Moon,	57753336	.	57753336	. . 0
— Apogee,	488203	.	488219	. . + 16
— Node,	232238	.	232226	. . — 12

	Surya Siddhanta.		Spurious Ārya Siddhanta.	
Mercury	17937060	.	17937020	diff. . — 40
Venus,	7022376	.	7022388	. . + 12
Mars,	2296832	.	2296824	. . — 8
Jupiter,	364220	.	364224	. . + 4
Saturn,	146568	.	146564	. . — 4

From this table it will readily appear, by the differences in the numbers, that the spurious *Ārya Siddhanta*, is placed several centuries anterior to the *Surya Siddhanta*, which work, however, is quoted by the former, as also Brahma Gupta and the *Brahma Siddhanta*. The date given to it is the year 3623 of the *Kali Yuga*, or A.D. 522, which is 16 years before the sideral form now in use was introduced. All these facts, I think, are sufficient, nowithstanding all the care that has been taken to prevent detection, to show it to be a mere modern imposition.

The first thing that strikes a person, on looking over this trifling work, is the order in which the planets are named, being the same as in the European or Mahomedan books of astronomy, and contrary to the universal practice of all other Hindu astronomers, who arrange the names of the planets in the same order, as they are expressed by the days of the week; that is, the sun, moon, Mars, Mercury, Jupiter, Venus, and Saturn. Whereas the writer of this spurious production, names them in the following order: the sun, moon, Mercury, Venus, Mars, Jupiter, and Saturn; or reversing it, by beginning with Saturn. From which circumstance I think it may be naturally inferred, that some European or Mahomedan tables had been employed in the framing of its numbers, or at least in correcting them, and perhaps by European hands. Indeed, for some years past, many things have appeared, as if written by Hindus, in which the assistance at least of Europeans appears conspicuous; such as the vaccine innoculation, Sir Isaac Newton's method of fluxions, &c. &c. It is certainly true, that as the Hindus become more and more acquainted with the European languages, writings, and arts and sciences, they are thereby enabled to give a view of the European sciences in Sanscrit, and father them on some of their own ancient sages, if they think, that by so doing, they will exalt themselves in the eyes of the world, by making them believe that their forefathers were the persons from whom the arts and sciences, even in their present improved state, originally sprung.

But to return. The next thing that attracts attention in the spurious *Ārya Siddhanta* is the commencing of the *Kalpa*, or grand period, with *Thursday*, instead of *Sunday*, which is entirely contrary to all the Hindu books that give regular systems. This, most probably, may be owing to an ignorance of the method of constructing a regular system; for to a person acquainted with the process, it is as easy to make Sunday the first day of the *Kalpa*, as any other day: but Sunday being the first day of the week, it is natural that it should be the first day of the *Kalpa*, and is so given in all the books of astronomy that I have seen, except the spurious *Ārya Siddhanta*. The *Kalpa*, by this book, begins at sunrise, 1986120000 years before the beginning of *Kali Yuga*. In this number of years there are 725447570625 days complete, which, divided by 7, give 103635367232 weeks, and one day over: therefore, as the *Kali Yuga* began with Friday, the *Kalpa* began on Thursday.

The system, in every point of view, is imperfect: it gives no revolutions for computing the places of the aphelia and nodes of the planets, nor any rule for determining the precession of the equinoxes for any particular period of time required; thus showing a want of knowledge in the requisites for forming a complete system.

But though he neither gives the revolutions of the nodes nor apsides, he gives the positions of the aphelia of the planets, which however are computed from the numbers given in the genuine *Ārya Siddhanta* (Section III.): this completely overturns the whole machinery, and shows it to be modern, in spite of all the art and cunning that had been employed in the forgery. Beside the decisive fact here mentioned, there are other circumstances in the book, that would also show that it is of a late date; for the author exhibits his knowledge in modern improvements. For instance, he gives the ratio, or proportion of the diameter of a circle to its circumference, as 20000 to 62832; that is, as 10000 to 31416, or 1 to 3.146; the multiplying it by 20000 was evidently to disguise it. This proportion of the diameter to the circumference, was totally unknown in the days of the genuine Āryabhatta, and is in fact the very same as given by Bhāskara Āchārya, in another form, in the *Lilāvati*.

It also states the diameter of the earth,	1050	yojans.
of the sun,	4410	
of the moon,	315	
of Venus,	65	
of Jupiter,	3220	
of Mercury,	2140	
of Saturn,	1615	
of Mars,	65	
of Meru,	1	
And the orbit of the wind,	3375	

with other particulars, of no interest or value to the astronomer. It contains about 20 small leaves, and goes by the name of the *Laghu*, or small *Ārya Siddhanta*. The genuine *Ārya Siddhanta* contains 40 leaves or more, exclusive of the part that is lost.

It would be needless to waste any more time in going over its contents: what has been shown must be perfectly sufficient to convince any man of common sense of its being a downright modern forgery, intended to supersede the genuine *Ārya Siddhanta*. There is, however, one passage in it which deserves to be noticed, as it will be of use hereafter in showing other forgeries to be brought forward. The passage relates to Brahma Gupta, whom he mentions in quoting the *Brahma Siddhanta*, and stands thus: —

ब्रह्मगुप्तः॥ नवेदनी यो जातककर्म यद्वतिं। नब्रह्मसिद्धांतमवानपि ग्रहान्॥
यात्राविवाहादिककार्यं सूचके। ज्ञेयोत्र मस्त्युः खलु तत् स्वरूपष्टक

He also quotes from the *Surya Siddhanta*; but in so doing there is no inconsistency; for the genuine Āryabhatta was long posterior to the times of either of these works. The principal forgery consists in the attempt to falsify the epoch of Āryabhatta, by throwing him back into antiquity, and in suppressing, for that purpose, the genuine *Ārya Siddhanta*.

PULISA SIDDHANTA.

There is another work, called the *Pulisa Siddhanta*, which is said to have been written in opposition to the doctrines contained in the spurious *Ārya Siddhanta*. It gives the same number of revolutions to the planets as the spurious *Ārya Siddhanta*, except for Mercury, which it makes 17937000, instead of 17937020; and for Jupiter 364220, instead of 364224, which, however, may arise from miscopying. The number of days given by Pulisa, as he is called, for a *Mahā Yuga*, or 4320000 years, is 1577917800, which divided by 4, gives 394479450 in 1080000 years; and these being divided by 7, give 56354207 weeks, and one day over: hence the system, in respect of commencement and cycles, is the same with the *Surya Siddhanta*. And as the length of the year is 365^{days} $15^{da.}$ 31′ 30″, it must commence at midnight, in order that, in reckoning from the beginning of the *Kali Yuga*, it may agree, at least nearly so, with the *Surya Siddhanta* in respect to the moon's place, &c. For if it was to begin at sunrise, then at the end of the year 3600 of the *Kali Yuga*, it would differ from the *Surya Siddhanta* by nearly six hours; and the error in the moon's place would be proportional to that difference, because the revolutions of the moon given in the *Pulisa Siddhanta* are the same with those given in the *Surya Siddhanta*. This Pulisa also follows the spurious *Ārya Siddhanta*, in adopting the same proportion of the diameter of a circle to its circumference, viz. 1 to 3.146, which he employs in computing the diameters of the orbits of the planets. Thus, the circumference of the orbit of Venus being 2664632 yojans, the diameter will be 848176, and the semidiameter 424088, as given in Bhattotpala's quotations: so that, though the *Pulisa Siddhanta* be feigned to be written in opposition to the spurious *Ārya Siddhanta*, it is only like one impostor attacking another, as by mutual consent, on some trivial point, but with a view of supporting him in that which is the main object of both — imposition in respect of time. Bhattotpala, or the person who assumes that name, whom I have shown to have lived about the middle of last century, I believe was the author of this *Pulisa Siddhanta;* for he seems to show the same partiality to it, in exhibiting its system, as Āryabhatta had done in respect to the *Parāsara Siddhanta,* which was doubtless his work.

The fact is, that literary forgeries are now so common in India, that we can hardly know what book is genuine, and what not: perhaps there is not one book in a hundred, nay, probably in a thousand, that is not a forgery, in some point of view or other; and even those that are allowed or supposed to be genuine, are found to be full of interpolations, to answer some particular ends: nor need we be surprised at all this, when we consider the facilities they have for forgeries, as well as their own general inclination and interest in following that profession; for to give the appearance of antiquity to their books and authors increases their value, at least in the eyes of some. Their universal propensity to forgeries, ever since the introduction of the modern system of astronomy and immense periods of years in A.D. 538, are but too well known to require any further elucidation than those already given. They are under no restraint of laws, human or divine, and subject to no punishment, even if detected in the most flagrant literary impositions.

THE SPURIOUS BRAHMA SIDDHANTA,

AND

BRAHMA SIDDHANTA SPHUTA.

We come now to notice another forgery—the *Brahma Siddhanta Sphuta**, the author of which I

* This spurious production was found by Mr. Colebrooke on the shelf of his library, as he himself declared, without knowing he had it. He could not know

believe I know. The object of this forgery was to throw Varāha Mihira, who lived about the time of Akber, back into antiquity, by placing him before the time of Brahma Gupta, who is supposed to have been the author of the *Brahma Siddhanta,* framed about the year A.D. 538, and the oldest of all the modern astronomical works now extant.

The *Brahma Siddhanta Sphuta,* is not, however, the only forgery that has been contrived to serve the like purpose; for it appears that a spurious or interpolated *Brahma Siddhanta* is made to quote Āryabhatta in several places. Now it is to be particularly remarked, that the quotations made are derived from the spurious *Ārya Siddhanta* above given; and yet the spurious *Ārya Siddhanta* quotes both Brahma Gupta and the *Brahma Siddhanta.* Here, then, is a downright contradiction, which is to be accounted for by the forger not having the spurious *Ārya Siddhanta* in his possession, and therefore could not foresee that it might ultimately detect him. The passages interpolated were to be found in the *Sarvabhauma,* or its commentary, and perhaps in other books.

Thus we see how Brahma Gupta, a person who lived long before Āryabhatta and Varāha Mihira, is made to quote them, for the purpose of throwing them back into antiquity. Pulisa is also introduced for the same purpose.—Now what does all this prove? It proves most certainly that the *Brahma Siddhanta* cited, or at least a part of it, is a complete forgery, probably framed, among many other books, during the last century by a junta of Brahmins, for the purpose of carrying on a regular systematic im-

---

that he had it, because it was purposely placed there for him to find, no doubt, by the person who framed it. More need not be said on the subject; this is sufficient.

position. Under this view of the case, we need not be surprised at agreements in quotations, respecting the *Brahma Siddhanta,* said to be found in different books; for these books themselves may have been forged, or interpolated to answer the purpose intended. We see a complete combination throughout, and therefore cannot, surely, place any reliance on assertions or quotations that are expressly contradicted by astronomical facts: and therefore, we must consider that Brahma Gupta, was either a mere modern of last century, or else that the books now attributed to him, are downright forgeries, either in part or the whole.

Mere agreement in quotations, nay, allowing the quotations to be just, and to have been actually in the original, will never make a work or works that are spurious genuine also; for it must be well known, that the practice of impostors is to interlard their productions with genuine passages, in order to give strength to the forgeries, and cause deception. The *Brahma Siddhanta* being shown to be a forgery, at least in part, from the quotations of the names of Varāha Mihira, Āryabhatta, Pulisa, and others (contained, as Mr. Colebrooke says, in the eleventh chapter of that work), any reference made to it to support the authenticity of the *Brahma Siddhanta Sphuta,* cannot alter the fact of that being also a forgery. The passage referred to by Mr. Colebrooke for this purpose is, however, a mere fiction, intended to convey an idea of the extraordinary antiquity of the Hindu modern systems. The passage is this:—"The computation of the planets taught by Brahma, which had become imperfect by great length of time, is propounded correct *(Sphuta)* by Brahma Gupta, son of Jishnu."

In the first place, it is acknowledged, on the face of the passage, that the computation of the places of the planets taught by Brahma, became imperfect by great length of time (that is, to give an appearance of antiquity to the system). In the second place, it states that Brahma Gupta propounds them as correct. Then who is the person that states them to have become incorrect? It could not be Brahma Gupta, since he propounds them as correct; for the passage does not say that he corrected them: consequently, since they are stated to be correct in the time of Brahma Gupta, he must be the identical framer of the system, and of the *Brahma Siddhanta* containing it, under the fictitious name of Brahma. But what is more remarkable is, that "the computation is propounded correct by Brahma Gupta, the son of Jishnu;" which evidently appears as if spoken or written by another person, and not by Brahma Gupta, who would naturally say, "is propounded correct by me."

Under all these circumstances, the passage is evidently a forgery, and was never written by Brahma Gupta: but whether it has any allusion to the *Brahma Siddhanta Sphuta*, under the colour of the word *Sphuta*, I will not pretend to say; but there is no doubt but that is also a forgery, as I have already stated.

The following is another passage quoted by Mr. Colebrooke from this *Brahma Siddhanta*, wherein Brahma Gupta is made to say: "I will refute the errors respecting the *Yugas*, and other matters, of those who, misled by ignorance, maintain things contrary to the *Brahma Siddhanta*." In the former passage, Brahma Gupta is made to speak in the third person, in this in the first person, which is rather inconsistent; but it does not seem that either should be by him. The latter appears as a passage by some person coming forward to support with all his might, a work already written, under the name of the *Brahma Siddhanta;* so that it could not be supposed to mean the work in which the passage is found, and of course then unfinished. These two passages alone, I think, would be sufficient to prove the forgery, without any other assistance; for they are both said to be taken from one and the same work, which I call the spurious *Brahma Siddhanta.*

Mr. Colebrooke seems to have been led into a belief, from the forgeries above mentioned, that ancient works contained improvements not to be found in more modern ones.

It is certainly possible, that an earlier writer may have an improvement in his work, not noticed by a later one who has not seen it, and therefore ignorant of the circumstance: but this cannot possibly be the case where a writer quotes another; for then he knows the whole of the work he cites, and cannot be ignorant of the improvement made: consequently, wherever such circumstances appear, they are sure indications of forgery.

Thus, in the spurious *Brahma Siddhanta,* Brahma Gupta is made to quote Āryabhatta, and also a part of his system. The *Ārya Siddhanta* quoted by him is the spurious one, in which the proportion of the diameter of a circle to its circumference is stated at 20000 to 62832. But Brahma Gupta, in his geometry, does not give this proportion, but states it as 1 to the square root of 10, or as 1 to 3.16227, which is a proof that Brahma Gupta, the author of the geometry under his name, never saw the spurious *Ārya Siddhanta,* though he is made to quote that work. The spurious *Brahma Siddhanta,* together with the spurious *Ārya Siddhanta,* are doubtless the productions of the last century, at farthest; but the *Brahma Siddhanta,* in the state in which it is quoted by Mr. Colebrooke, may be even of the present century.

I shall now introduce a passage which Mr. Colebrooke has brought forward, by way of supporting his opinion respecting the positions of the colures. He says: "The passage in which this author, Brahma Gupta, denies the precession of the colures, as well as the comment of his scholiast on it, being material to the present argument, they are here subjoined in a literal version."*

"*The very fewest hours of night occur at the end of Mithuna* (Gemini), *and the seasons are governed by the sun's motion. Therefore the pair of solstices appears to be stationary, by the evidence of a pair of eyes.*" That is to say, according to this passage of Brahma Gupta, the solstices are always fixed to the beginning of Cancer and Capricorn, which is strictly true, and are so now. Brahma Gupta, by this passage, did not say there was no precession in respect of the fixed stars: all that he meant and declared was, that the solstice was fixed to the beginning of Cancer and Capricorn in the moveable sphere, and had no reference whatever to the sideral sphere, which is fixed, and the signs of which, beginning with the Lunar Asterism *Aswinī,* have no names, but are expressed numerically; consequently there was not the slightest ground for misunderstanding the meaning of Brahma Gupta, or attempting to give it a turn he never intended. He further adds:

The commentator (Prithudaca Swami) says: "*What*

---

* Notes and Illustrations, xxxvi.

is said by Vishnu Chandra, *at the beginning of the chapter on the Yuga of the solstice* [*the revolutions, though the asterisms are here in* 4320000000 *years a hundred and eighty nine thousand four hundred and eleven* (189411), *which is termed a Yuga (revolution) of the solstice, as of old admitted by* Brahma, Arca, *and the rest*], *is wrong;* for *the very fewest hours of night to us occur when the sun s place is at the end of Mithuna (Gemini), and of course the very utmost hours of day are at the same period. From that limitary point, the sun's progress regulates the seasons, namely, the cold season (Sisira), and the rest, comprising two months each, reckoned from Macara (Capricorn): therefore what has been said concerning the motion of the limitary point is wrong, being contradicted by actual observation of days and nights.*" This is precisely the same as stated by Brahma Gupta, and has no relation whatever to the precession of the equinoxes in respect of the fixed stars, which is Vishnu Chandra's meaning. They are both right according to the sense in which they themselves meant.

Then comes an interpolation contradicting Prithudaca Swami, which I have every reason to believe, from the nature of the questions put to me on the subject of the colures by a certain astrologer, was surreptitiously inserted by him, by which he makes the commentator contradict what he said just before; for there is no other person mentioned.

" The objection, however, is not valid; for *now* the greatest decrease and increase of night and day do not happen when the sun is in the end of *Mithuna*" (Gemini). * By this artful interpolation, he thought to overturn the opinion I had given him,

* Notes and Illustrations, xxxvi.

which was, that the solstices were *now,* and at all times, in the beginning of Cancer and Capricorn. He has shown by it, not only his own ignorance of the Hindu astronomy, but also his propensity to forgery; for there is hardly a work on astronomy that would not expose the imposition. The matter of fact is this:—The Hindu astronomers, as I have stated in another place, employ two spheres, the sideral or fixed, which commences from the beginning of the Lunar Asterism *Aswinī:* the other sphere is moveable, and is precisely the same with what is called the European or Tropical sphere, in which the signs begin from the vernal equinoctial point. In the former, the signs are merely numerically expressed; but in the latter, to distinguish them from the former, they are named as in Europe, Aries, Taurus, Gemini, &c.; therefore they can never be confounded, except through ignorance, inadvertence, or for the sake of imposition, as done in the above quoted passage, where the third Hindu sign is falsely called *Mithuna* (Gemini), as will be fully and satisfactorily proved in the next section.

With respect to Vishnu Chandra's number of revolutions of the equinoxes in 4320000000 years, being stated at 189411, it is certainly wrong, when applied to any of the known *Kalpas.* Mr. Colebrooke at one time was of the same opinion*: but in a note which he has added to his paper, in the twelfth volume of the Asiatic Researches, on the precession of the equinoxes, he has altered that opinion, and states the number as right†; which, however, on the principles of the Hindu astronomy, it cannot possibly be so, because, if tried with the years now

---

* As. Res. Vol. XII. p. 215.

† Note after p. 250.

elapsed of any of the known *Kalpas*, it will not give the quantity of the precession for the present time, which is the only mode of proving whether the number be true or otherwise; for if it does not answer the purpose for which it was intended, of course it must be considered as either incorrect, or an imposition.

In the second section (page 131), I have shown that the number of years elapsed of the system of Brahma to the year 4900 of the *Kali Yuga*, was 1972948900, and that the precession of the equinoxes then amounted to 0ˢ 21° 9′ 34″. If we now try Vishnu Chandra's number 189411 for the same period, we shall have by the formula, $\frac{189411 \times 1972948900}{4320000000} =$ (86504 Revol.) 2ˢ 18° 40′ 29, differing from the truth by upwards of 57°; and in like manner it is found to differ in all the other known systems.

But Brahma Gupta is made to speak of this Vishnu Chandra (*Brahma Siddhanta*, C. 11 and 14), stating, that he was the author of the *Vasisht'ha Siddhanta*, and that he took the mean motions of the sun and moon, with the lunar apogee and nodes, and other specified particulars, from Āryabhatta: * then, if so, the system of Vishnu Chandra must be the same with that of Āryabhatta; for the numbers must be always framed to answer the system of years, and will not agree with any other. The object here, as may be easily seen, is to throw Āryabhatta back into antiquity; for if Vishnu Chandra borrows his system from Āryabhatta, and Brahma Gupta mentions the circumstance, then it is evident both the one and the other must have been anterior to him, that is to say, provided it is not a mere fiction

* As. Res. Vol. XII. Note after p. 250.

invented for the purpose. We shall now try Vishnu Chandra's number above given, with the system of Āryabhatta, in which the time elapsed to the year 4900 of the *Kali Yuga*=1969924900 years, from which the precession of the equinoxes will be $=\frac{189411 \times 1969924900}{4320000000}=$ (86371 Revol.) $7^s\ 17°\ 6'$ which differs greatly from the truth: therefore this cannot be the system supposed to be employed. Let us, therefore, try the system of the spurious *Ārya Siddhanta.* By this system, the number of years elapsed of the *Kalpa* to the year 4900 of the *Kali Yuga*=1986124900: therefore the precession will be $=\frac{189411 \times 1986124900}{4320000000}=$ (87081 Revol.) $11^s\ 1°\ 57'$, and which also differs widely from the truth: therefore, I say the number is an imposition. But even if the number had been right, it would not have altered the fact of the passage being an imposition, in respect of Brahma Gupta who is thus made to speak of persons that lived many centuries after his own time.

All this, and perhaps a great deal more not yet brought to light, is, I am satisfied, the fabrication of the astrologer already alluded to. He offered his services to me before he was in the employ of Mr. Colebrooke; but when he told me that his profession was book-making, and that he could forge any book whatever, to answer any purpose that might be required, I replied, I wanted no forged books — that there were too many of that description already — that I was extremely glad he was so candid, and must decline his services in any way whatever. In the course of the conversation that passed, he made himself acquainted with Mr. Colebrooke's opinions that were in opposition to mine, which it seems he carefully treasured up in his mind. He went directly to Mr. Colebrooke's from my house, and there got into immediate employ, as he himself afterwards informed me. This will serve to explain the circumstance of the forged book (the *Brahma Siddhanta Sphuta*), being found by Mr. Colebrooke on the shelf in his library, without knowing that he had it; as also the various forgeries of names and quotations in the spurious *Brahma Siddhanta*, made up on purpose, to throw the persons named back into antiquity to answer the end in view; but in so doing, he was detected and foiled by the very books of the authors themselves, which showed the times in which they lived and wrote, beyond the power of forgery to pervert or contradict. More, I think, need not be said; and I hope this will put an end to the subject for ever, particularly as the forgeries are incontestibly proved, independent of all other considerations and circumstances whatever.

SECTION VI.

Self-defence the object of this Section — The notions of Mr. Colebrooke respecting the positions of the stars at the general epoch, as indicating the age of the works in which they are found, inconsistent with real facts, being given in books of all ages — Mr. Colebrooke's notions respecting the names Aries, Taurus, &c. being applied to the signs of the Hindu sphere, incorrect — Proved to belong exclusively to the signs of the Tropical sphere, by tables and passages in modern Hindu books — Passage from Brahma Gupta *to the same effect — Passage from* Varāha Mihira *to the same effect — Another from the* Tatwa-chintamani, *containing a computation of the sun's place, reckoned both from Aswinī and Aries, to the same effect — A translation of a passage in* Bhattotpala's *Commentary to the same effect — The translation by Mr. Colebrooke himself, but not published or noticed by him — Nor the other facts stated — Mr. Colebrooke notices the heliacal rising of Canopus at Ujein, when the sun was 7° short of Virgo, mentioned by* Varāha Mihira, *but does not tell us the time to which it refers — Notices other risings, but without reference to time — Mistaken with respect to the time of rising of Canopus in the time of* Pārāsara *— A passage from Garga explained — Three periods of Canopus's heliacal rising — The 8th and 15th of Āswina, and 8th of Kārtika, mistranslated by Mr. Colebrooke — The true meaning given — The time to which they refer explained in a note.*

Having in the foregoing sections given a sufficient outline of the nature of the modern astronomical systems of the Hindus, and exposed the various practices employed for imposing on the world the pretended antiquity of their books and writers, I should now most willingly wish to drop my pen, particularly as I believe I have omitted nothing that could in any way be conducive to the perfect understanding of the subject. It appears, however, that there is something more yet to be done, however unpleasant the task may be; and that is, to come forward in my own defence.

There are two points on which Mr. Colebrooke seems to have laid great stress, in his endeavours to support the antiquity of the *Surya Siddhanta* and Varāha Mihira against the result of my calculations; which two points, though already noticed, we shall here endeavour to show are totally inapplicable, and therefore mere delusions, without the slightest foundation.

One of the points relates to the longitudes of the stars, reckoned from the beginning of the Hindu sphere, commencing with the Lunar Asterism *Aswinī*, as contained in the tables of the Lunar Mansions in different books. These longitudes, from the very nature of the subject, must in every case be the same, or nearly the same, whether given by an astronomer who lived a thousand years ago, or by one who lived only fifty years ago: from the point from which the longitudes are reckoned being fixed to the commencement of the Hindu sphere when the precession was nothing, the longitudes of the stars reckoned from that point, must of necessity be always the same, though given at different ages by different astronomers, except so far as one may be more or less accurate in his computation than another, which, however, can never point out the difference of time. Thus, some Hindu astronomer about the year A.D. 538, observed the longitude of Cor Leonis, from the beginning of *Aswinī*, in the commencement of the Hindu sphere, to be $4^s\ 9°$; and another, about the middle of last century, made it also $4^s\ 9°$. Now I should be glad to know how this is to point out the difference of time between the two astronomers, or when they respectively lived. I say it is impossible; but even supposing the latter astronomer had made it $4^s\ 8°$, or

4^s 10°, the difference, in that case, could only arise from a greater or less degree of accuracy in the observation made, and had nothing whatever to do with time. Hence it most clearly follows, that the longitudes of the stars, reckoned from the commencement of the Hindu sphere, can never point out the time when any astronomer lived, or any book in which they are given was written or composed. But, notwithstanding the clearness of this fact, and the soundness of the foundation on which it stands, Mr. Colebrooke has endeavoured to prove the antiquity of the *Surya Siddhanta* from the longitudes of the stars given in that work, reckoning from the commencement of the Hindu sphere. In the *Brahma Siddhanta*, the longitude of the star Cor Leonis is reckoned at 4^s 9° from the commencement of the Hindu sphere: in the *Surya Siddhanta* it is also 4^s 9°: in the *Varahi Sanhita* it is 4^s 9°: in the *Siddhanta Siromani* it is 4^s 9°: in the *Siddhanta Sarvabhauma*, a still more modern book, it is also 4^s 9°: and in the *Grahu Laghava*, another modern work, it is also 4^s 9°: all exhibiting the same longitude, though composed or written at different dates, and by different persons. The reason of this must be obvious. A certain point is fixed on by all the Hindu astronomers, from which they compute the precession of the equinoxes: at that point they also give the positions of the stars in the Lunar Asterisms, or what they suppose they were at that time; and this makes all Hindu writers, let their respective times be what it may, agree nearly with each other; for the positions so given, have no reference whatever to the age or time of the writer or his book, but merely to the common epoch to which they all refer; that is to say, the point of time when there was no precession, and the beginning of *Aswinī* was cut by the equinoctial colure. Therefore tables computed for this point, for the sake of uniformity and convenience, are found in books of various dates, or no dates expressed, without having the slightest reference to the time of any of them. Hence I say that the whole of Mr. Colebrooke's notions on this point are altogether unfounded.

Indeed the facts against such ideas are incontestible. The observation made by Vasisht'ha on the star Canopus, who found it in 3^s, or the beginning of Cancer, which is the same as given in the *Surya Siddhanta*, would fully prove this; but the positions of the planets given by that work, prove that it was even of a much later date than the observation on Canopus by Vasisht'ha. Mr. Colebrooke, however, is not disposed to admit the correctness of this mode of determining the antiquity of astronomical books by the positions of the planets, except where it suits his purpose. He saw it was sufficiently correct in the case of the *Brahma Siddhanta*; but he would not admit it to be so with respect to the *Surya Siddhanta*: and why? Because Varāha Mihira, whom he imagined had lived thirteen or fourteen hundred years ago, had mentioned the *Surya Siddhanta*: therefore he was determined to adopt a new mode (by the longitudes of the fixed stars from the beginning of *Aswinī*), for determining the age of the *Surya Siddhanta*, which mode, if exact, ought to determine the ages of all other books also; but the truth is, it neither gives the age of the *Surya Siddhanta*, nor of any other work whatever, as may be easily seen from the explanation above given. But to put this in a still stronger light, suppose we designate the books above mentioned by the letters of the alphabet, and call them *a*, *b*, *c*, *d*, and *e*, stating that they were of different ages, and that each of them made the longitude of Cor Leonis from the beginning of the Hindu sphere $4^s\ 9^\circ$, and then ask Mr. Colebrooke, or any other person, which was the oldest: it is clear that he could not tell. He would perhaps say, let me know the name of the book, and I will then tell you. But there is no magic in the name of the book; and if the method could not determine the question, without knowing the name of the book, it is no method whatever; it is a downright delusion. Mr. Colebrooke might perhaps say, that as they gave the same longitude, they might have been borrowed from each other, the more modern from the older; but this is not the case, and even if it was so, it would not be solving the question; for the *Surya Siddhanta* would then stand in the same predicament, as the author of it might also have borrowed, and therefore its real time would still be totally unknown. But we do not determine the antiquity of books by the most ancient facts or expressions they may contain, but by the most modern. The longitude of Canopus, as given by Vasisht'ha, in the beginning of Cancer, is also in the *Surya Siddhanta;* and this is more modern than the epoch of the commencement of the modern astronomy, to which all the tables of the Lunar Asterisms refer, whether they be found in a book written last year, or a thousand years ago: and the positions of the planets given by the *Surya Siddhanta,* show that it is still more modern than even the observation of Vasisht'ha. And to crown all this, the very individual on whose account the age of the *Surya Siddhanta* was to be perverted and twisted about, turns out to have been contemporary with the emperor Akber. Had the Hindu astronomers given us the longitudes of the stars from the vernal equinoctial point, at the times they respectively wrote, we should from thence be able to determine the times in which they lived; but this they took particular care not to do, knowing well what would be the consequence: on the contrary, they reduce the longitudes reckoned from the vernal equinoctial point, whatever they may find them, to what they would appear to be from the beginning of the Hindu sphere, or *Aswinī*, at the general epoch, when the vernal equinox was supposed to coincide with it, which being a fixed point, the longitudes must be the same, or nearly so, by all; and by that means puts it out of the power of the most acute astronomer that ever was, to determine the times when they wrote, or the ages of the books in which they were given; which shows the complete fallacy of Mr. Colebrooke's method, if it can be so called.

We shall now proceed to Mr. Colebrooke's other point, on which he seems to have laid so much stress, in supporting the supposed antiquity of Varāha Mihira, &c.

Varāha Mihira states, that one solstice is in the beginning of Cancer, and the other in that of Capricorn. Hence Mr. Colebrooke says, that he must have lived about thirteen hundred years ago, because he has assumed that the names Aries, Taurus, Gemini, Cancer, &c. belong to the signs of the Hindu sideral sphere, beginning from *Aswinī*, and therefore concludes, that the solstitial points were not in the beginning of Cancer and Capricorn for these last thirteen hundred years. But in so doing, Mr. Colebrooke has drawn a most incorrect conclusion; for the real matter of fact is, that the names Aries,

Taurus, &c. as repeatedly stated, belong to the signs of the tropical sphere, beginning from the vernal equinoctial point, and not in any manner whatever to the signs of the Hindu sideral sphere, which we shall now proceed to prove beyond dispute.

I have already mentioned, that the tropical sphere was received by the Hindus from the west, I believe about the second or third century of the Christian era, with the names of the signs, the same as they are still in use in Europe: that on introducing the sideral sphere now in use in A.D. 538, the tropical sphere was still retained for certain purposes; and in order that no confusion should arise, the names Aries, Taurus, &c. were retained for the signs, commencing from the vernal equinoctial point, while those of the sideral sphere, commencing from *Aswinī*, were represented numerically only, by which means all confusion was avoided. Now I will show that the same rule is still followed by the modern Hindus; and for which purpose I beg leave to introduce the following table of the sun's right ascension and declination, as now in use, for the end of every sign, reckoning from the beginning of *Mesha* (Aries), or the vernal equinoctial point.

PERPETUAL TABLE OF THE SUN'S RIGHT ASCENSION AND DECLINATION.

Names of the signs.		Sun's right ascension at the end of each sign.		Difference of right ascension.	Declination.
Mesha, ...	Aries,	1670′	27° 50′	1670′	11° 43′ N.
Vṛisha, ...	Taurus,	3465	57 45	1795	20 38
Mithuna, ...	Gemini,	5400	90 0	1935	24 0
Karkata, ...	Cancer,	7335	122 15	1935	20 38
Sinha, ...	Leo,	9130	.152 10	1795	11 43
Kanyā, ...	Virgo,	10800	180 0	1670	0 0
Tula, ...	Libra,	12470	207 50	1670	11 43 S.
Vrischika, ...	Scorpio,	14265	237 45	1795	20 38
Dhanus, ...	Sagittarius,	16200	270 0	1935	24 0
Makara, ...	Capricornus,	18135	302 15	1935	20 38
Kumbha, ...	Aquarius,	19930	332 10	1795	11 43
Mina, ...	Pisces,	21600	360 0	1670	0 0

In the first column we have the names of the signs in *Sanscrit*, with the same translated. The second and third columns contain the sun's right ascension to the end of every sign. The fourth column contains the differences between the right ascensions in the second column; and the fifth contains the sun's declination at the end of each sign, corresponding to the right ascension. Those who may not have it in their power to consult Hindu books or tables, may refer to Mr. S. Davies's papers in the second volume of the Asiatic Researches, pages 271 and 272, where they will find the names of the signs, the sun's right ascension, corresponding to the end of each sign, separately taken, which corresponds with the fourth column above, and the sun's declination for the same points. All these, and a great deal more, will be found in a modern Hindu work, called the Tables of *Makaranda*, to which Mr. Davies refers.

By reference to the table, it will appear that the sun's right ascension at the end of *Mithuna* (Gemini), according to the Hindus, is always 5400′ or 90°, and that the declination of the sun in the same point is 24° N. that the sun's right ascension at the end of *Dhanus* (Sagittarius), is 16200′ or 270°, and declination 24° S. Now as the table is perpetual, it follows that the signs named, are not those of the Hindu sphere beginning with *Aswinī*, but those of the tropical sphere, beginning from the mean vernal equinoctial point: — the Hindu astronomers, by thus inserting the names of the signs in their tables, prevent any possibility of mistake. Brahma Gupta says, that "the very shortest hours of night occur at the end of *Mithuna* (Gemini), and the seasons are governed by the sun's motion. Therefore the pair of solstices appear to be stationary by the evidence of a pair of eyes." And is not this the case by the above table, which is now in use? Are not the solstitial points always in the beginning of Cancer and Capricorn? Brahma Gupta did not refer to the Hindu sphere, for the names of the signs do not belong to it; and therefore his own simple expression ought to have been sufficient to point out what he meant, without perverting it to another purpose; nor is it possible that any real astronomer could misunderstand him.

Varāha Mihira is equally explicit, and his meaning clear and unequivocal. He says: "At present one solstice is in the beginning of *Karkata* (Cancer), and the other in that of *Makara*" (Capricorn); and he again says: "The sun, by turning without having reached *Makara* (Capricorn), destroys the south and the east; by turning back without having reached *Karkata* (Cancer), the north and east. By turning when he has just passed the summer *solstitial point,* he makes wealth secure, and grain abundant, since he moves thus according to nature; but the sun, by moving unnaturally, excites terror." By this passage of Varāha Mihira, the solstices were always in the beginning of Cancer and Capricorn. Are they not so now? and are they not so by the table above given now in use? Where is, then, the foundation of the inference drawn from thence by Mr. Colebrooke, that he lived thirteen or fourteen hundred years ago? The foundation does not exist in truth, and the whole error arises from Mr. Colebrooke assuming that the signs named belonged to the sideral sphere; but this could not be, from the nature of the expression used, which referred to the tropical sphere only, and could not be mistaken.

I will now adduce another proof from the *Tatwa-chintamani,* a modern work by Lakshmidas, who gives examples for calculating the sun's right ascensions, declinations, &c. in all of which he takes care to distinguish the tropical from the sideral sphere.

In one of the examples he states the sun's mean longitude in the Hindu sideral sphere, that is, from the beginning of *Aswinī,* at $11^{s}\ 2^{\circ}\ 9'\ 6''$

He then adds the precession or difference of the spheres, 0 16 10 59

The sum he calls *Mina* (Pisces), 18 20 5

Can any thing be more clear and decisive than this example, to show that the names of the signs are reserved alone for the tropical sphere? The sun's longitude in the Hindu sphere is simply expressed by figures, without the name of any sign being mentioned; whereas in the tropical sphere it is particularly marked with the name of the sign, the degrees, minutes, and seconds, being put down in figures. To say that the sun's mean longitude at one and the same moment of time was $11^{s}\ 2^{\circ}\ 9'\ 6''$ and $11^{s}\ 18^{\circ}\ 20'\ 5''$ would appear inconsistent; and to say that it was in Pisces $2^{\circ}\ 9'\ 6''$ and Pisces $18^{\circ}\ 20'\ 5''$ at the same time, would be equally so: there was, therefore, no better way of distinguishing the spheres than by affixing the name of the sign to that to which it properly and originally belonged. The sun's longitude being thus expressed, serves as the foundation for computing the right ascension and declination for that point, as also the time of sun rising and setting, length of the day and night, with other particulars that may be required, all depending on the tropical sphere.

I shall give one example more from another

modern writer, under the name of Bhattotpala, whom I have already mentioned, and who appears to have lived in the last century, but thrown back, by the imposition of Dr. Hunter's pundit while at *Ujein*, and no other authority, to the year A.D. 968. The passage relates to the method of determining the times of the solstices, and is thus: — "The observations of the solstices, or sun's motion between the solstices, is to be made at sunrise by the intersection of a distant mark; for the sun having reached the beginning of *Capricorn*, moves daily towards the north, and being arrived at the beginning of Cancer, moves daily towards the south. Therefore, marking the sun's place at sunrise or sunset by some distant object, as a tree, &c. examine the sun's place again next day, to ascertain whether the sun's motion or declination has stopped or not; and the observations may be continued for seven days after the sun's arrival at the beginning of the sign, by computation, to determine whether the computed true place agree, precede or recede. Or in a large arch having delineated on smooth ground a circle marked relatively to the quarters, erect in its centre a gnomon, then at the equinox at sunrise and sunset the shadow falls on the lines east and west. As long as the sun advances towards the end of *Gemini*, the shade at sunrise continues to move south of the line east and west, and the same at sunset: it then moves south until it reach the end of *Virgo*, and the beginning of *Libra* the shadow falls on the line east and west. From that time the shadow advances daily north till it reaches *Capricorn*, and then recedes daily south to the end of *Pisces*."

This passage is as clear as possibly can be, in showing that the signs named belong to the tropical sphere, and not to the Hindu. Indeed there is not an instance that I know where the names of the signs, as Aries, Taurus, &c. have been applied to the signs of the Hindu sphere: if such, however, has occurred, it must arise either from inadvertence, ignorance, or for the purpose of imposition, as in the interlopation mentioned in the last section, (p. 191), where the commentator is made to say, that the greatest decrease and increase of night and day do not appear *now,* when the sun's place is at the end of *Mithuna* (Gemini), because the interpolator would wish to support the ideas that the names of the signs belong to the Hindu sphere; but which, by the facts we have shown above, is completely refuted: and therefore Mr. Colebrooke's other strong point, founded on the assumption that the names of the signs are those of the Hindu sphere, beginning from *Aswinī,* completely falls to the ground, as totally contrary to the Hindu astronomy and to facts.

Indeed Mr. Colebrooke had ample means of correcting his ideas, if it was his wish so to do; for in fact the above passage from Bhattotpala is an actual translation by Mr. Colebrooke, and written in his own hand on the margin of the original, which was borrowed for me, and from which I copied it. Will it not, therefore, appear very singular that he should bring forward such an interpolation to support his opinion, while at the same time he kept back this passage in Bhattotpala, which he himself translated? But this is not all: he has a copy of the *Tatwa-chintamani,* made from the one in my possession, from which the example of the sun's longitude above given was taken. He might also have seen the tables of the sun's right ascensions, declinations, &c. for each sign in different Hindu books of astronomy, and must have seen the table given by Mr. Davies already alluded to. No excuse can, therefore, exist in holding an opinion so entirely contrary to facts and to the Hindu astronomy, which carefully assigns, in every instance, the names Aries, Taurus, &c. to the signs of the tropical sphere only.

Mr. Colebrooke takes notice of the sun's longitude being within 7° of Virgo at the time of the heliacal rising of Canopus at *Ujein,* according to the testimony of Varāha Mihira*, which, had he determined the time to which it corresponded, he would not only have seen that Varāha Mihira was a modern, but that his works, interpolations, and assertions on which he so much relied, were mere impositions, of very modern date. The trick was first played off on the emperor Akber, and ever since continued and supported by all the ingenuity that Brahminical cunning and imposition could suggest or invent. Mr. Colebrooke notices the rules given in the *Bhāsvati* and the *Grahalāghava* for the rising of Canopus†; but he does not tell us to what time or times they refer; but, what is still more singular, he seems to understand them as rules for the visible rising of Canopus.‡ Now, according to my own ideas of astronomy, this is impossible, because the sun's longitudes, by the rules, come out less than in the time of Varāha Mihira; whereas they should have been greater, in proportion to the times posterior to him. The rules, in fact, give the cosmical risings of Canopus, and not the heliacal, as is evident also from the authors themselves, who only state, that

---

* As. Res. Vol. IX. p. 355 and 356. † As. Res. Vol. IX. p. 356.
‡ As. Res. Vol. IX. p. 357.

when the sun is in the longitude given by the rule, then the star Canopus rises with the sun, and not a syllable about its being visible: but even if they had said so, it would be of no use, as it would appear from the statement of Varāha Mihira, which must be our guide, to be an imposition.

Mr. Colebrooke also notices the heliacal rising of Canopus in the time of Pārāsara.* This was when the sun's longitude was in the beginning of the Lunar Asterism *Hastā,* and at the same period it set heliacally when the sun was in the beginning of *Rohinī;* from which Mr. Colebrooke concludes, that " the right ascension of Canopus must have been in his time not less than 100°, reckoned from the beginning of *Mesha* (Aries), and the star rising cosmically, became visible in the oblique sphere at the distance of 60° from the sun, and disappeared, setting achronically when within that distance."

Mr. Colebrooke has here evidently misapplied the name of the sign *Mesha* (Aries), as he has done in many other instances; for it is not the name of the first sign of the Hindu sideral sphere, but the first reckoned from the vernal equinoctial point, as has been fully proved.

In the fifth section of the first part, I have shown the real time of Pārāsara to have been 575 B.C., in which year Canopus rose heliacally, when the sun was in the beginning of the Lunar Asterism *Hastā,* and at the same time distant from the vernal equinox $4^{s}\,25°\,10'\,5''$ in the latitude of Delhi. In that year the right ascension of Canopus from the vernal equinoctial point, was found to be $81°\,43'\,25''$, and its longitude $2^{s}\,8°\,47'$. The vernal equinoctial point was

* As. Res. Vol. IX. p. 357.

then about 14° 50′ to the east of the beginning of the fixed *Aswinī.*

Mr. Colebrooke does not say what method he adopted, or the data he employed in his calculation; but it would appear from what he says, that he assumed the colure to have been in the middle of *Asleshā* in the time of Pārāsara. If Mr. Colebrooke meant by this, the middle of the fixed Lunar Asterism *Asleshā,* which I suppose he did, it is incorrect; for Pārāsara was contemporary with Yudhisht'hira and Garga, and the latter wrote his *Sanhita* in the year 548 B.C. therefore the colure could not then be in the point assumed, nor at any later period than the year 1192 B.C. which was even 247 years before the time of Rāma. But if Mr. Colebrooke meant the tropical or anastral *Asleshā,* the assumption would be correct; for the ancients had two spheres, the one fixed, the other moveable; that is, the sideral and tropical; the same in fact, as the moderns still employ, and for the like purposes, though differently divided. The moderns have their fixed constellations and moveable signs: the latter are always reckoned from the vernal equinoctial point, or the intersection of the equator and ecliptic. The ancients had in like manner their fixed and moveable Lunar Mansions, which were called by the same names, and the latter always began from the winter solstice: hence one of the solstitial points was always in the beginning of the moveable Lunar Mansion *Sravisthā,* and the other in the middle of the moveable *Asleshā.* So that Garga is right when he says: "When the *sun* returns, not having reached *Dhanisht'hā* (i.e. *Sravishthā*), in the northern solstice, or not having reached the middle of *Asleshā* in the southern, then let a man feel great apprehension of danger."* And Pārāsara says: "When having reached the end of *Sravanā* in the northern path, or half of *Asleshā* in the southern, he still advances, it is a cause of great fear."* Thus in fact expressing the same thing that Varāha Mihira had done, in respect of the solstitial points being always in the beginning of Cancer and Capricorn, from which the expressions are of equal import in either case: but instead of collecting or conceiving the natural and true meaning of the Hindu writers, it was supposed that they were ignorant of the precession of the equinoxes, than which nothing could be more incorrect; for the precession was not only known to Varāha Mihira, but also to Garga and Pārāsara, and was even known long before their times. If such unaccountable mistakes could have been made in respect of the tropical signs, or those beginning from the vernal equinox, which I proved above to be the fact, can we wonder at similar ones being made in respect of the moveable Lunar Mansions, which always begin from the winter solstice, and have the same names as the fixed, but recede or fall back by reason of the precession?

Mr. Colebrooke also states, that Bhattotpala cites from the *Pancha Siddhantica* a rule of computation analogous to that which is given in the *Bhāsvati;* and remarks, that three periods of Canopus' heliacal rising are observed, viz. the 8th and 15th of *Āswina*, and 8th of *Cartica.*†

The rule here alluded to, as cited from the *Pancha Siddhantica*, I have already given in the last section, which shows most clearly that it is a forgery

---

* As. Res. Vol. V. p. 397. † As. Res. Vol. IX. p. 356.

of the last century only, and to which period its rule refers.

With respect to the three periods noticed, viz. the 8th and 15th of *Āswina* and 8th of *Cartica*, there is a mistake in the translation of the words, the original being *Astamī* and *Panchadasī*. These names of the days, as is well known to all the Hindus, invariably refer to the moon's age, and never to the day of the month: therefore the passage should have been translated, "the 8th and 15th lunar day of *Āswina*, and 8th of the moon of *Cartica*." The former reading is totally irreconcilable to facts; but the latter is easily understood by every person that has a sufficient knowledge of the Hindu astronomy.*

* The Hindus, as I have already observed, employ two spheres, the sideral and tropical; and to the signs in both they have corresponding months, which bear the same name: the moment the sun enters a sign, that instant the month also begins; so that by knowing the name of the month, we also know the sign. The signs which here are designated by *Āswina* and *Cartica*, are Virgo and Libra in the tropical sphere, because the heliacal risings of Canopus are reckoned according to that sphere. The moon is named after two different, and I may say directly opposite methods: in one it is named from the month or sign in which the new moon begins; in the other from the month or sign in which it terminates: the last is the method here meant. Thus, if there is a new moon in *Āswina* or Virgo, and the end falls in Libra (*Cartica*), it is called the moon of *Cartica*; and in like manner, when the moon ends in Virgo, it is called the moon of Virgo, or *Āswina*. Now to apply all this to the solution of the problem, it will be seen that the 8th day of the moon of Virgo can never fall later than about the 8th or 9th degree of Virgo, which is therefore one limit. On the other hand, it may be seen that the 8th day of the moon of Libra can never begin earlier than the beginning of Virgo, because its end must terminate in Libra: therefore the 8th day of the moon, in this case, falls also on the same point, viz. on the 8th or 9th degree of Virgo, which determines that Canopus rose heliacally in some part of India when the sun was between the 8th or 9th degree of Virgo. To ascertain the place this could happen at, take a celestial globe, bring Canopus to the eastern horizon, then elevate the globe until it is found that the 8th or 9th degree of Virgo is 10 degrees below the horizon, while at the same time Canopus is just on it, then the degrees of elevation of the pole will show the latitude, which will be found to be that of Delhi, and that the time to which it refers is the last century; which is a further confirmation that Bhattotpala is a modern, and the *Pancha Siddhantica* a forgery. The 15th day of the moon of *Āswina* (Virgo) would of course fall, in some years, on the 8th or 9th day of the month of *Āswina*;

Thus I have shown that Mr. Colebrooke had the same means and facts before him that I had, for investigating the truth or falsehood of those passages and assertions that are intended to throw back into antiquity the time of Varāha Mihira and others: on this, however, I shall offer no comment, but leave it to others. All that I can say is, that I have strictly done my duty, notwithstanding all the difficulties I have had to encounter, and the opposition thrown in my way. My sole object has been the investigation of truth; and little I expected at setting out, that I should find nothing but inveterate enmity as the reward of my labours—but so it is. It is by the investigation of truth, and the exposure of Brahminical impositions, which can only be done through the means of astronomy, that the labours of those who are laudably endeavouring to introduce true religion and morality among the Hindus can have their true and beneficial effect. So long as the impositions and falsehoods contained in the Hindu books, which the common people are made to believe are the productions of their ancient sages, are suffered to remain unexposed, little progress can be expected to be made: but let the veil be undrawn, uncover the impositions by true and rational investigation, and the cloud of error will of itself disappear; and then they will be not only more ready, but willing to adopt and receive the word of truth. The time, however, is now come that I must relinquish these pursuits. Ill health for some years past, with an enfeebled constitution, from a long residence

---

so that the three periods, so called, are nót very distant ones. The method of calling the moon by the name of the month in which it ends is very ancient, though at present little used. In Bengal, and many other parts of India, the moon is named from the month in which it begins.

in a warm climate, having been between forty and fifty years in India, obliges me to lay down my pen, and to desist from all further investigations: indeed it has been with a great deal of difficulty that I have brought this essay to a close, in the state it is in. If my health had permitted, it might have been made more perfect and full; but on this account I have been obliged to curtail it, and leave out many things that might be useful. However, though thus narrowed, I believe the astronomer and man of science will still find all that he may require, or that is absolutely necessary, for forming a just idea of the Hindu astronomy, and its antiquity.

THE APPENDIX,

CONTAINING

I.—*Hindu Tables of Equations, &c. for calculating the true Heliocentric and Geocentric Places of the Planets, &c.*

II.—*Remarks on the Chinese Astronomy, proving, from their Lunar Mansions, that the Science is much more modern among them than is generally believed. The names of their Constellations are added, with the Stars in each.*

III.—*Translations of certain Hieroglyphics, which hitherto have been called (though erroneously) the Zodiacs of Dendera in Egypt.*

No I.

HINDU TABLES OF EQUATIONS, &c.

FOR CALCULATING

THE TRUE PLACES OF THE SUN AND MOON;

ALSO THE

HELIOCENTRIC AND GEOCENTRIC PLACES OF THE PLANETS.

HAVING been requested by some friends to add to my work the Tables of Equations, &c. used by the Hindus in computing the true places of the sun and moon, and the heliocentric and geocentric longitudes of the planets, I feel a pleasure in complying with the request, by the insertion of the accompanying Tables, which I believe have never before been published, those of the sun and moon excepted. For the better understanding the tables, I have added examples under each, so that nothing more need be said here by way of explanation.

SUN.

EQUATION OF THE ORBIT.
Argument. Sun's mean anomaly.

Degrees.	+ 6ˢ / − 0ˢ	+ 7ˢ / − 1ˢ	+ 8ˢ / − 2ˢ	Degrees.
0	0° 0′ 0″	1° 6′ 02″	1°53′ 25″	30
1	0 2 20	1 8 00	1 54 30	29
2	0 4 40	1 9 57	1 55 34	28
3	0 7 00	1 11 57	1 56 35	27
4	0 9 19	1 13 47	1 57 34	26
5	0 11 37	1 15 40	1 58 34	25
6	0 13 56	1 17 32	1 59 30	24
7	0 16 15	1 19 23	2 0 23	23
8	0 18 33	1 21 11	2 1 14	22
9	0 20 51	1 22 57	2 2 04	21
10	0 23 7	1 24 42	2 2 51	20
11	0 25 23	1 26 26	2 3 35	19
12	0 27 39	1 28 07	2 4 17	18
13	0 29 55	1 29 46	2 4 57	17
14	0 32 10	1 31 23	2 5 35	16
15	0 34 27	1 32 58	2 6 12	15
16	0 36 37	1 34 32	2 6 45	14
17	0 38 39	1 36 04	2 7 17	13
18	0 41 01	1 37 35	2 7 45	12
19	0 43 12	1 39 06	2 8 12	11
20	0 45 22	1 40 36	2 8 35	10
21	0 47 31	1 42 03	2 8 58	9
22	0 49 39	1 43 26	2 9 18	8
23	0 51 47	1 44 45	2 9 36	7
24	0 53 53	1 46 02	2 9 51	6
25	0 55 57	1 47 17	2 10 03	5
26	0 58 01	1 48 33	2 10 13	4
27	1 00 02	1 49 47	2 10 20	3
28	1 2 53	1 51 00	2 10 27	2
29	1 4 03	1 52 22	2 10 31	1
30	1 6 02	1 53 25	2 10 32	0
Degrees.	− 5ˢ / + 11ˢ	− 4ˢ / + 10ˢ	− 3ˢ / + 9ˢ	Degrees.

Use of the Table.

Supp. sun's mean long. is	11ˢ	2°	9′	6″
Subtract the apogee	2	17	56	38
Sun's mean anomaly	8	14	12	28
The equation for which is +		2	5	42
Sun's mean longitude ..	11	2	9	6
Sun's true longitude	11	4	14	48

MOON.

EQUATION OF THE ORBIT.
Argument. Moon's mean anomaly.

Degrees.	+ 6ˢ / − 0ˢ	+ 7ˢ / − 1ˢ	+ 8ˢ / − 2ˢ	Degrees.
0	0° 0′ 00″	2° 32′ 2″	4°22′ 29″	30
1	0 5 20	2 36 37	4 25 26	29
2	0 10 40	2 41 11	4 27 36	28
3	0 16 00	2 45 36	4 29 59	27
4	0 21 19	2 49 58	4 32 19	26
5	0 26 36	2 54 20	4 34 37	25
6	0 31 54	2 58 39	4 36 47	24
7	0 37 12	3 2 54	4 38 54	23
8	0 42 29	3 7 5	4 40 54	22
9	0 47 44	3 11 12	4 42 50	21
10	0 52 18	3 15 16	4 44 40	20
11	0 58 11	3 19 18	4 46 24	19
12	1 03 23	3 23 24	4 48 5	18
13	1 08 40	3 27 26	4 49 38	17
14	1 13 45	3 30 54	4 51 9	16
15	1 18 53	3 34 39	4 52 53	15
16	1 24 00	3 38 21	4 53 54	14
17	1 29 05	3 41 58	4 55 6	13
18	1 34 9	3 45 32	4 56 15	12
19	1 39 10	3 48 59	4 57 17	11
20	1 44 9	3 52 24	4 58 13	10
21	1 49 17	3 55 46	4 59 6	9
22	1 54 3	3 59 2	4 59 53	8
23	1 58 3	4 2 13	5 0 27	7
24	2 3 47	4 5 18	5 1 8	6
25	2 8 35	4 8 18	5 1 40	5
26	2 13 22	4 11 16	5 2 3	4
27	2 18 6	4 14 11	5 2 20	3
28	2 22 47	4 17 0	5 2 36	2
29	2 27 35	4 19 46	5 2 44	1
30	2 32 2	4 22 29	5 2 48	0
Degrees.	− 5ˢ / + 11ˢ	− 4ˢ / + 10ˢ	− 3ˢ / + 9ˢ	Degrees.

Use of the Table.

Supp. moon's mean long. =	11ˢ	8°	56′	17″
The moon's apogee	2	9	38	0
Moon's mean anomaly ..	8	29	18	17
The equation for which is +		5	2	45
Moon's mean longitude	11	8	56	17
Moon's true longitude ..	11	13	59	2

MERCURY.

EQUATION OF THE SUN'S LONGITUDE.
Arg. Sun's mean long.—Aphelion of Mer.

Degrees.	— 0s	— 1s	— 2s	— 3s	— 4s	— 5s	Degrees.
0	0′	134′	232′	267′	236′	139′	30
1	5	138	234	268	234	135	29
2	10	142	236	268	232	131	28
3	14	145	238	267	229	126	27
4	19	149	240	267	227	122	26
5	24	153	242	267	224	117	25
6	29	157	244	267	222	113	24
7	34	160	246	266	220	108	23
8	38	164	248	266	217	104	22
9	43	167	249	265	213	99	21
10	47	171	250	264	211	95	20
11	52	174	251	264	207	90	19
12	56	178	252	263	204	85	18
13	61	181	253	262	201	80	17
14	65	185	255	261	198	75	16
15	70	191	257	260	195	70	15
16	74	194	258	259	192	65	14
17	79	197	259	258	188	60	13
18	84	200	260	257	185	55	12
19	88	203	261	256	181	51	11
20	92	206	262	254	178	46	10
21	96	209	263	252	175	41	9
22	101	212	264	251	171	36	8
23	105	214	265	249	167	31	7
24	109	217	265	248	163	26	6
25	114	220	266	246	159	21	5
26	118	222	266	244	155	15	4
27	122	225	267	242	151	10	3
28	126	227	267	240	147	5	2
29	130	230	267	238	143	1	1
30	134	232	267	236	139	0	0
Degrees.	11s +	10s +	9s +	8s +	7s +	6s +	Degrees.

Use of the Table.

Suppose sun's long.....	11s	2°	9′	16″
Mercury's aphelion subt.	7	14	54	40
Remain..............	3	17	14	26
The equation for which is —		4	17	45
Subt. from sun's long. =	11	2	9	6
Remain sun's equat. long.	10	27	51	21

Note. The true heliocentric longitude of Mercury is not used by the Hindu astronomers.

THE ELONGATION.
Arg. Mercury's long.—Sun's equated long.

Degrees.	+ 0s	+ 1s	+ 2s	+ 3s	+ 4s	+ 5s	Degrees.
0	0′	478′	902′	1208′	1240′	906′	30
1	16	494	914	1216	1235	884	29
2	32	508	926	1222	1230	860	28
3	48	522	938	1228	1225	838	27
4	64	536	950	1234	1220	808	26
5	82	552	964	1240	1218	790	25
6	98	568	976	1246	1216	762	24
7	114	582	988	1252	1214	736	23
8	130	594	1000	1256	1212	708	22
9	142	612	1010	1260	1210	684	21
10	162	628	1020	1266	1208	656	20
11	178	642	1032	1270	1204	626	19
12	194	656	1044	1274	1194	596	18
13	210	670	1054	1278	1182	566	17
14	226	686	1064	1282	1172	536	16
15	242	700	1076	1284	1160	506	15
16	258	714	1086	1286	1148	476	14
17	274	728	1096	1288	1136	444	13
18	290	742	1106	1290	1122	410	12
19	306	756	1116	1292	1106	378	11
20	324	770	1123	1290	1092	346	10
21	338	784	1134	1290	1078	302	9
22	352	796	1144	1290	1062	278	8
23	368	810	1152	1290	1044	244	7
24	384	824	1160	1290	1026	210	6
25	400	838	1170	1282	1008	174	5
26	414	848	1178	1278	990	140	4
27	430	862	1184	1262	970	104	3
28	446	876	1192	1254	948	70	2
29	462	890	1200	1246	928	34	1
30	478	902	1208	1240	906	0	0
Degrees.	11s —	10s —	9s —	8s —	7s —	6s —	Degrees.

Use of the Table.

Suppose Mercury's long.	5s	20°	51′	15″
Subt. sun's equated long.	10	27	51	21
Remains the commutation	6	22	59	54
The elong. for which is.. —		12	16	0
Sun's equated long.	10	27	51	21
Geocentric long. of Merc.	10	15	35	21

VENUS.

EQUATION OF THE SUN'S LONGITUDE.
Arg Sun's mean long—Venus's Aphelion.

Degrees.	− 0^s	− 1^s	− 2^s	− 3^s	− 4^s	− 5^s	Degrees.
0	0′	56′	91′	105′	92′	54′	30
1	2	57	92	105	91	52	29
2	4	59	93	105	90	51	28
3	6	60	94	105	89	49	27
4	8	62	95	105	88	47	26
5	10	63	95	105	87	46	25
6	12	65	96	105	86	44	24
7	15	66	97	105	85	42	23
8	17	67	98	105	84	39	22
9	19	68	99	104	83	38	21
10	21	70	99	104	82	37	20
11	23	71	100	104	81	35	19
12	25	72	100	103	79	33	18
13	27	73	101	103	78	31	17
14	29	74	101	103	77	29	16
15	30	75	102	102	76	28	15
16	32	76	102	101	74	26	14
17	34	78	103	101	73	24	13
18	36	79	103	101	72	22	12
19	37	80	103	100	70	20	11
20	39	81	104	99	68	18	10
21	41	82	104	99	67	16	9
22	43	83	104	98	66	14	8
23	44	84	105	97	65	12	7
24	46	85	105	97	63	10	6
25	48	86	105	95	62	8	5
26	49	87	105	95	61	6	4
27	51	88	105	94	60	4	3
28	53	89	105	93	58	2	2
29	54	90	105	93	56	1	1
30	56	91	105	92	54	0	0
Degrees.	11^s +	10^s +	2^s +	8^s +	7^s +	6^s +	Degrees.

Use of the Table.

Suppose sun's long.	11^s	2°	9′	6
Venus's aphelion subtract	2	21	17	10
Remain	8	10	51	56
The equation for which is	−	1	40	0
Sun's longitude	11	2	9	6
Sun's equated longitude	11	3	49	6

Note. The true heliocentric longitude of Venus is not used.

THE ELONGATION.
Arg. Venus's long.—Sun's equated long.

Degrees.	+ 0^s	+ 1^s	+ 2^s	+ 3^s	+ 4^s	+ 5^s	Degrees.
0	0′	754′	1484′	2152′	2666′	2656′	30
1	26	778	1506	2172	2678	2636	29
2	50	802	1532	2194	2690	2608	28
3	76	828	1554	2214	2702	2578	27
4	100	852	1576	2222	2712	2544	26
5	126	876	1600	2252	2722	2510	25
6	150	900	1624	2270	2732	2470	24
7	174	928	1646	2290	2740	2430	23
8	202	954	1670	2310	2748	2384	22
9	228	976	1692	2330	2756	2336	21
10	254	1000	1714	2350	2782	2280	20
11	278	1026	1738	2368	2768	2218	19
12	302	1050	1760	2384	2772	2156	18
13	328	1074	1784	2402	2778	2090	17
14	352	1098	1804	2422	2782	2014	16
15	378	1124	1824	2440	2784	1934	15
16	404	1148	1848	2456	2784	1836	14
17	428	1172	1872	2474	2784	1746	13
18	452	1196	1892	2490	2784	1646	12
19	478	1220	1916	2510	2782	1546	11
20	504	1244	1938	2524	2778	1430	10
21	530	1268	1962	2540	2774	1316	9
22	554	1292	1982	2556	2768	1194	8
23	580	1316	2004	2570	2760	1060	7
24	604	1340	2024	2586	2750	922	6
25	630	1364	2044	2602	2740	778	5
26	654	1388	2068	2616	2726	630	4
27	678	1406	2088	2628	2702	480	3
28	704	1436	2110	2642	2694	320	2
29	728	1460	2120	2654	2676	160	1
30	754	1484	2152	2666	2656	0	0
Degrees.	11^s −	10^s −	9^s −	8^s −	7^s −	6^s −	Degrees.

Use of the Table.

Suppose Venus's long. ..	0^s	16°	30′	11″
Subtract Sun's equat. long.	11	3	49	6
Remains the commutation	1	12	41	5
The elongation for which is	− 17	39	34	
Sun's equated longitude ..	11	3	49	6
Venus's geocentric long.	11	21	28	40

MARS.

EQUATION OF THE ORBIT.
Arg. Mars's mean long — the Aphelion.

Degrees.	– 0s	– 1s	– 2s	– 3s	– 4s	– 5s	Degrees.
0	0′	320′	570′	689′	629′	381′	30
1	11	330	576	690	624	370	29
2	23	340	583	691	618	359	28
3	33	350	589	691	613	348	27
4	45	359	595	692	607	336	26
5	56	369	600	692	601	325	25
6	67	378	606	692	594	313	24
7	78	387	611	692	587	301	23
8	89	396	617	692	580	289	22
9	100	405	622	691	573	277	21
10	111	415	627	690	566	265	20
11	122	423	632	690	559	253	19
12	133	432	636	689	551	240	18
13	145	440	641	687	543	228	17
14	155	449	645	685	537	215	16
15	165	458	649	683	527	202	15
16	176	466	653	680	519	189	14
17	187	475	657	679	510	176	13
18	197	483	660	676	501	162	12
19	208	492	663	673	492	149	11
20	219	498	667	670	483	136	10
21	232	506	670	667	474	123	9
22	239	514	673	664	464	109	8
23	250	521	676	661	454	96	7
24	261	528	678	657	444	82	6
25	270	536	680	652	434	69	5
26	280	543	682	648	423	55	4
27	291	550	684	644	413	42	3
28	301	557	686	639	403	27	2
29	311	563	687	634	392	14	1
30	320	570	689	629	381	0	0
Degrees.	11s +	10s +	9s +	8s +	7s +	6s +	Degrees.

SEMIPARALLAX OF THE ORB.
Arg. Sun's longitude — Mars's longitude.

Degrees.	+ 0s	+ 1s	+ 2s	+ 3s	+ 4s	+ 5s	Degrees.
0	0′	352′	687′	984′	1186′	1096′	30
1	12	363	698	994	1190	1081	29
2	24	375	709	1002	1193	1065	28
3	36	387	720	1010	1196	1048	27
4	47	398	730	1019	1199	1030	26
5	59	410	740	1026	1202	1010	25
6	71	421	751	1035	1204	990	24
7	83	432	761	1043	1205	968	23
8	95	444	773	1051	1206	944	22
9	106	455	781	1059	1208	919	21
10	118	467	792	1067	1208	892	20
11	130	478	803	1075	1208	864	19
12	142	489	813	1082	1208	833	18
13	154	500	823	1089	1207	801	17
14	164	512	833	1096	1206	767	16
15	177	523	842	1103	1204	731	15
16	189	534	852	1110	1202	694	14
17	201	545	862	1117	1199	655	13
18	213	556	872	1124	1195	613	12
19	224	568	882	1130	1191	559	11
20	237	579	892	1136	1187	525	10
21	248	590	901	1141	1183	474	9
22	259	601	910	1147	1175	430	8
23	271	612	920	1153	1168	380	7
24	283	623	930	1158	1160	329	6
25	295	634	939	1164	1152	277	5
26	306	644	948	1169	1141	223	4
27	318	655	957	1173	1131	168	3
28	329	666	966	1178	1121	112	2
29	341	677	975	1182	1109	56	1
30	352	687	984	1186	1096	0	0
Degrees.	11s –	10s –	9s –	8s –	7s –	6s –	Degrees.

Use of the Table.

Suppose Mars's mean long.		2s	7°	35′	18″
The aphelion		4	8	24	57
Anomaly		9	29	10	21
The equation for which is	+		9	35	0
Mars's mean longitude ..		2	7	35	18
Mars's true helioc. long...		2	17	10	18

Use of the Table.

Sun's mean longitude ..		11s	2°	9′	6″
Mars's true long. subt ..		2	17	10	18
The commutation		8	14	58	48
The semiparallax ..	–		18	23	0
Its double	–	1	6	46	0
Mars's true helioc. long.		2	17	10	18
Mars's geocentric long...		1	10	24	18

JUPITER.

EQUATION OF THE ORBIT.
Arg. Jupiter's mean long — the Aphelion.

Degrees.	— 0s	— 1s	— 2s	— 3s	— 4s	— 5s	Degrees.
0	0′	149′	260′	305′	271′	161′	30
1	5	153	262	306	269	156	29
2	10	158	265	306	266	151	28
3	15	162	267	306	263	147	27
4	21	167	270	306	261	142	26
5	26	171	272	305	259	137	25
6	31	175	275	305	256	132	24
7	38	179	277	305	252	127	23
8	44	184	279	304	249	123	22
9	48	188	281	304	246	116	21
10	52	192	283	303	242	111	20
11	57	196	285	303	239	105	19
12	62	200	287	302	235	100	18
13	67	203	289	301	232	94	17
14	72	207	291	300	229	89	16
15	77	211	292	299	225	83	15
16	82	215	294	298	221	78	14
17	87	218	295	297	217	72	13
18	92	221	296	295	213	66	12
19	97	225	297	294	209	61	11
20	102	229	299	292	205	56	10
21	107	232	300	290	201	50	9
22	112	236	301	289	197	44	8
23	117	239	301	287	193	39	7
24	121	242	302	285	188	34	6
25	126	245	303	283	184	29	5
26	131	248	304	281	180	24	4
27	135	251	304	279	175	19	3
28	140	254	305	276	170	12	2
29	144	257	305	274	166	6	1
30	149	260	305	271	161	0	0
Degrees.	11s +	10s +	9s +	8s +	7s +	6s +	Degrees.

Use of the Table.

Supp. Jupiter's mean long.	10s	25°	47′	6″
Subtr. long. of the aphelion	5	22	35	16
Remains anomaly	5	3	11	50
The equation for which is —		2	26	0
Mean longitude of Jupiter	10	25	47	6
True heliocentric long...	10	23	21	6

SEMIPARALLAX OF THE ORB.
Arg. Sun's longitude — Jupiter's long.

Degrees.	+ 0s	+ 1s	+ 2s	+ 3s	+ 4s	+ 5s	Degrees.
0	0′	145′	268′	339′	326′	203′	30
1	5	149	271	340	324	198	29
2	10	154	274	342	321	191	28
3	15	158	278	343	318	186	27
4	19	163	280	344	316	180	26
5	24	167	284	344	313	173	25
6	29	172	287	344	309	167	24
7	34	176	290	345	306	161	23
8	39	180	293	345	304	154	22
9	44	184	296	346	301	148	21
10	48	189	298	346	297	141	20
11	53	193	301	345	294	135	19
12	59	197	304	345	290	127	18
13	64	202	307	345	285	121	17
14	68	206	310	345	282	114	16
15	73	210	312	345	278	107	15
16	78	214	314	344	274	100	14
17	83	219	316	344	269	96	13
18	87	223	318	343	265	86	12
19	92	228	320	342	261	76	11
20	97	231	323	342	256	67	10
21	102	234	325	341	251	58	9
22	107	238	327	339	246	50	8
23	111	242	329	338	241	43	7
24	116	246	331	336	236	35	6
25	121	249	332	335	231	28	5
26	125	253	334	333	225	21	4
27	130	257	335	331	220	14	3
28	135	260	336	330	215	7	2
29	140	264	338	328	209	1	1
30	145	268	339	326	203	0	0
Degrees.	11s —	10s —	9s —	8s —	7s —	6s —	Degrees.

Use of the Table.

Sun's longitude........	11s	2°	9′	6″
Jupiter's true longitude..	10	23	21	6
The commutation	0	8	48	0
The semipar. for which is +			42	0
Its double .... = +		1	24	0
True longitude of Jupiter	10	23	21	6
Geocentric longitude....	10	24	45	6

SATURN.

EQUATION OF THE ORBIT.
Arg. Saturn's mean long—the Aphelion.

Degrees.	− 0s	− 1s	− 2s	− 3s	− 4s	− 5s	Degrees.
0	0′	219′	385′	459′	411′	245′	30
1	8	225	390	459	407	238	29
2	15	232	393	459	403	231	28
3	23	238	396	459	399	223	27
4	30	245	400	460	395	215	26
5	38	251	404	460	391	208	25
6	46	257	408	459	387	201	24
7	53	263	411	459	383	193	23
8	61	270	414	459	379	185	22
9	68	276	418	458	374	177	21
10	76	282	422	457	369	169	20
11	83	288	425	457	364	161	19
12	91	294	428	456	358	153	18
13	98	300	431	454	353	146	17
14	106	305	433	453	348	137	16
15	113	311	438	451	342	129	15
16	120	316	441	449	336	120	14
17	128	322	443	448	330	112	13
18	135	327	444	446	324	104	12
19	143	333	445	444	318	96	11
20	150	338	447	441	312	87	10
21	157	343	448	438	306	78	9
22	164	348	450	435	300	69	8
23	171	353	452	432	294	61	7
24	178	358	453	429	288	52	6
25	185	362	454	425	282	44	5
26	193	367	455	423	274	35	4
27	199	371	456	420	267	27	3
28	206	376	457	417	260	9	2
29	213	380	458	414	253	1	1
30	219	385	459	411	245	0	0
Degrees.	11s +	10s +	9s +	8s +	7s +	6s +	Degrees.

Use of the Table.

Supp. Saturn's mean long.	1s 4°12′56″
Saturn's aphelion	8 20 54 27
Mean anomaly	4 13 18 29
The equation for which is	− 5 51 27
Saturn's mean longitude	1 4 12 56
Saturn's true helioc. long.	0 28 21 29

SEMIPARALLAX OF THE ORB.
Arg. Sun's long—Saturn's true long.

Degrees.	+ 0s	+ 1s	+ 2s	+ 3s	+ 4s	+ 5s	Degrees.
0	0′	86′	156′	190′	174′	105′	30
1	3	88	158	190	172	101	29
2	6	91	159	191	171	98	28
3	9	94	161	191	170	95	27
4	12	96	162	191	168	91	26
5	14	98	164	191	166	88	25
6	17	101	166	191	164	85	24
7	21	103	168	191	162	81	23
8	24	106	169	191	160	78	22
9	27	109	170	191	158	75	21
10	29	111	172	190	156	73	20
11	32	113	173	190	154	71	19
12	35	116	174	190	152	68	18
13	38	118	176	190	150	65	17
14	40	121	177	189	147	61	16
15	44	123	178	189	145	58	15
16	47	126	180	188	142	55	14
17	51	128	181	187	140	51	13
18	54	130	182	187	138	47	12
19	56	133	183	187	135	43	11
20	59	135	184	186	132	40	10
21	62	137	185	185	130	36	9
22	65	140	186	183	127	32	8
23	68	143	187	182	124	29	7
24	71	144	187	181	121	25	6
25	73	145	188	180	119	22	5
26	76	149	188	179	116	18	4
27	79	152	189	178	113	15	3
28	81	153	189	176	110	11	2
29	83	155	189	175	108	5	1
30	86	156	190	174	105	0	0
Degrees.	11s −	10s −	9s −	8s −	7s −	6s −	Degrees.

Use of the Table.

Saturn's true longitude..	0s 28°21′56″
Subtract from sun's long.	11 2 9 6
Commutation..........	10 3 47 37
The semipar. for which is	− 2 29 39
Its double is	− 4 59 18
Saturn's true longitude..	0 28 21 29
Saturn's geocentric long.	0 23 22 11

THE preceding examples exhibit the usual mode of the Hindu astronomers in general. There are, however, some who pretend to greater accuracy, and go through ten or a dozen operations to get the geocentric place of a planet; the method of which I will now show from the *Tatwa Chintāmani* of Lakshmī Dāsa, a commentary on the *Siddhānta Siromani*.

Let the geocentric longitude of Saturn be required, and his mean

Heliocentric place, as in the last example = ...		1ˢ	4°	12′	56″
Place of the aphelion subtract		8	20	54	27
Remains the mean anomaly		4	13	18	29
Equation of the orbit for which is....................	–		5	47	50
Saturn's mean longitude............................		1	4	12	56
Saturn's true longitude once equated		0	28	25	6
Which taken from the sun's mean longitude.............		11	2	9	6
Remains the first commutation		10	3	44	0
The parallax of the orb for which is..................	–		4	58	26
Which taken from his longitude once equated...........		0	28	25	6
Leaves the geocentric place once equated...............		0	23	36	40

Second Operation.

From the geocentric place once equated		0	23	36	40
Subtract place of his aphelion		8	20	54	27
Remains the second anomaly..........................		4	2	42	13
The equation of the orbit for which is	–		6	40	52
Saturn's mean longitude..............................		1	4	12	56
Saturn's heliocentric longitude twice equated		0	27	32	4
Which subtract from the sun's mean longitude............		11	2	9	6
Remains the second commutation		10	4	37	2
The parallax of the orb for which is....................	–		4	57	0
Saturn's heliocentric longitude twice equated		0	27	32	4
Saturn's geocentric longitude twice equated.............		0	22	35	4

Third Operation.

From Saturn's geocentric place twice equated		0	22	35	4
Subtract place of his aphelion		8	20	54	27
Remains third anomaly		4	1	40	37
The equation of the orbit for which is	–		6	46	57
Saturn s mean longitude.............................		1	4	12	56
Saturn's heliocentric longitude three times equated		0	27	25	59
Which subtract from the sun's longitude		11	2	9	6
Remains third commutation		10	4	43	7
Parallax of the orb for which is.......................	–	0	4	54	21
Saturn's longitude thrice equated		0	27	25	59
Saturn's geocentric longitude thrice equated.............		0	22	31	38

Fourth Operation.

Saturn's geocentric place thrice equated		0	22	31	38
Subtract place of his aphelion		8	20	54	27

Remains fourth anomaly		4ˢ	1°	37′	11″
The equation of the orbit for which is	–		6	47	12
Saturn's mean longitude		1	4	12	56
Saturn's heliocentric longitude four times equated		0	27	25	44
Subtract the same from the sun's mean longitude		11	2	9	6
Remains the fourth commutation		10	4	43	22
Parallax of the orb for which is	–		4	54	20
Saturn's fourth equated longitude		0	27	25	44
Saturn's geocentric longitude four times equated		0	22	31	24

Fifth Operation.

Saturn's geocentric longitude four times equated		0	22	31	24
Subtract his aphelion as before		8	20	54	27
Remains the fifth anomaly		4	1	36	57
The equation of the orbit for which is	–		6	47	12
Saturn's mean longitude		1	4	12	56
Saturn's heliocentric longitude five times equated		0	27	25	44

Which coming out the same as in the fourth operation, the calculation terminates, and the geocentric and heliocentric longitudes of Saturn remain as in the fourth operation.

The same method also serves for Jupiter; but Mars requires a different one. Thus for Mars:—

Suppose the mean longitude of Mars	=	2ˢ	7°	35′	18″
Place of his aphelion subtract		4	8	24	57
Remains the anomaly		9	29	10	21
Equation of the orbit for which is	+		9	46	16
Its half	+		4	53	8
Mean longitude, add		2	7	35	18
The sum is Mars' longitude once equated		2	12	28	26
Which taken from the sun's mean longitude		11	2	9	6
Remains the first commutation		8	19	40	40
Semiparallax of the orb for which is	–		18	28	43
Which taken from the longitude once equated		2	12	28	26
Leaves Mars' geocentric longitude once equated		1	23	59	43

Second Operation.

Mars' geocentric longitude once equated		1	23	59	43
Subtract his aphelion (corrected once)		4	6	43	35
Anomaly		9	17	16	42
Equation of the orbit for which is	+		10	41	37
Mars' mean longitude		2	7	35	18
Mars' heliocentric longitude twice equated		2	18	16	55
Which subtract from the sun's mean longitude		11	2	9	6
Remains second commutation		8	13	52	11
Parallax of the orb corresponding	–	1	8	19	7
Mars' longitude twice equated		2	18	16	55
Mars' geocentric longitude twice equated		1	9	57	48

Third Operation.

Mars' geocentric longitude twice equated		1	9	57	48
Subtract the aphelion (twice corrected)		4	5	47	50
Anomaly		9	4	9	58

The equation for which is	+	•	5°	35′	18″
Mars's mean longitude		2	7	35	18
Mars' heliocentric longitude thrice equated		2	13	10	36
Which subtract from the sun's mean longitude		11	2	9	6
Remains the third commutation		8	18	58	30
Semiparallax of the orb corresponding	—	0	18	33	34
Mars' longitude thrice equated		2	13	10	36
Mars' geocentric longitude thrice equated		1	24	37	2

Fourth Operation.

Mars' geocentric longitude thrice equated		1	24	37	2
Subtract the aphelion (thrice corrected)		4	6	36	44
Anomaly		9	18	0	18
Equation of the orbit for which is	+		10	39	8
Mars' mean longitude		2	7	35	18
Mars' heliocentric longitude four times equated		2	18	14	26
Which subtract from the sun's mean longitude		11	2	9	6
Remains fourth commutation		8	13	54	40
Parallax of the orb corresponding	—	1	8	18	29
Mars' longitude four times equated		2	18	14	26
Mars' geocentric longitude four times equated		1	9	55	57

Fifth Operation.

Mars' geocentric longitude four times equated		1	9	55	57
Subtract the aphelion four times corrected		4	1	48	14
Anomaly		9	8	7	43
Half the equation of the orbit	+		5	35	19
Mars' mean longitude		2	7	35	18
Mars' heliocentric longitude five times equated		2	13	10	37

Which being the same with that found in the third operation, puts an end to the calculation, as all the rest comes out the same; therefore the heliocentric longitude of Mars is 2ˢ 13° 10′ 37″, and the geocentric longitude 1ˢ 9° 55′ 57″.

To find the Elongation and geocentric Longitude of Mercury.

Suppose the sun's place or mean longitude	=	11ˢ	2°	9′	6″
Subtract place of Mercury's aphelion		7	14	54	40
Remains the anomaly		3	17	14	36
The equation for which is	—		5	46	59
Sun's mean longitude		11	2	9	6
Sun's first equated longitude		10	26	22	7
Which taken from the mean longitude of Mercury		5	20	51	15
Remains first commutation		6	24	29	8
The elongation for which is	—		12	50	12
Sun's longitude once equated		10	26	22	7
Mercury's geocentric longitude once equated		10	13	31	55

Second Operation.

Mercury's geocentric longitude once equated		10	13	31	55
Mercury's aphelion subtract		7	14	54	40
Remains anomaly		2	28	37	15
The equation for which is	—		6	3	15

		s	°	′	″
Sun's mean longitude		11	2	9	6
Sun's longitude twice equated		10	26	5	51
Which taken from Mercury's mean longitude		5	20	51	15
Leaves second commutation		6	24	45	24
The elongation for which is	–		12	56	49
Sun's longitude twice equated		10	26	5	31
Mercury's geocentric longitude twice equated		10	13	8	42

Third Operation.

Mercury's geocentric longitude twice equated		10	13	8	42
Mercury's aphelion, subtract		7	14	54	40
Anomaly		2	28	14	2
The equation for which is	–		6	3	10
Sun's mean longitude		11	2	9	6
Sun's longitude thrice equated		10	26	5	56
Which taken from Mercury's mean longitude		5	20	51	15
Leaves the third commutation		6	24	45	19
The elongation for which is	–		12	57	6
Sun's longitude thrice equated		10	26	5	56
Mercury's geocentric longitude thrice equated		10	13	8	50

Fourth Operation.

Mercury's geocentric longitude thrice equated		10	13	8	50
Mercury's aphelion, subtract		7	14	54	40
Anomaly		2	28	14	10
The equation for which is	–		6	3	10

Which being the same as in the third operation, all the rest will be the same also: therefore the geocentric place of Mercury is found to be 10ˢ 13° 8′ 50″, and the elongation 12° 57′ 6″. Venus's geocentric longitude and elongation is found in the same way.

To find the mean heliocentric longitude from the true geocentric longitude, all the requisite data being given. This is the reverse of the former operations. An example will be sufficient.

Required the mean heliocentric longitude of Saturn, from

		s	°	′	″
His true geocentric longitude found above		0	22	31	24
Subtract the same from the sun's mean longitude		11	2	9	6
Leaves first commutation		10	9	37	42
Parallax of the orb for which is, with a contrary sign	+		4	33	59
Geocentric longitude, add		0	22	31	24
Sum		0	27	5	23
Subtract place of the aphelion		8	20	54	27
Anomaly		4	6	10	56
Equation of the orbit for which is, with a contrary sign	+		6	25	59
Add		0	27	5	23
The sum is the mean heliocentric longitude		1	3	31	22

Second Operation.

Sun's mean longitude		11	2	9	6
Subtract mean longitude last found		1	3	31	22

Second commutation		9^s	28°	37′	44″
Parallax of the orb for which, with a contrary sign	+		5	17	18
Add geocentric place as before		0	22	31	24
The sum is		0	27	48	42
Subtract the aphelion		8	20	54	27
Anomaly		4	6	54	15
The equation for which is, with a contrary sign	+		6	22	28
Add		0	27	48	42
The sum is the mean heliocentric longitude		1	4	11	10

Third Operation.

Sun's mean longitude		11	2	9	6
Subtract mean longitude last found		1	4	11	10
Third commutation		9	27	57	56
Parallax of the orb for which is, with a contrary sign	+		5	19	25
Geocentric longitude, add		0	22	31	24
The sum is		0	27	50	49
Subtract place of the aphelion		8	20	54	27
Anomaly		4	6	56	22
Equation of the orbit for which is, with a contrary sign	+	0	6	22	27
Add		0	27	50	49
The sum is the mean heliocentric longitude		1	4	13	16

Fourth Operation.

Sun's longitude		11	2	9	6
Subtract the mean heliocentric longitude		1	4	13	16
Fourth commutation		9	27	55	50
Parallax of the orb for which is, with a contrary sign	+		5	19	42
Add geocentric longitude		0	22	31	24
The sum is		0	27	51	6
Subtract place of the aphelion		8	20	54	27
Remains the anomaly		4	6	56	39
Equation of the orbit for which is, with a contrary sign	+		6	22	17
Add		0	27	51	6
The sum is the mean heliocentric longitude		1	4	13	23

Fifth Operation.

Sun's mean longitude		11	2	9	6
Subtract mean longitude last found		1	4	13	23
Fifth commutation		9	27	55	43
Parallax of the orb for which is, with a contrary sign	+		5	19	42

Which being the same as in the fourth operation, puts an end to any further calculation: so that 1^s 4° 13′ 23″ is the mean heliocentric longitude required, differing only about 27″ from the original in page 225.

Table of the Circumferences, Semidiameters, and Eccentricities of the Orbits of the Planets, in Yojans, according to* Dādā Bhāi, *a Commentator on the Surya Siddhanta.*

Planets.	Orbits.		Eccentricities.		
	Circumfer.	*Semidiamet.*	*Least.*	*Greatest.*	*Mean.*
Sun....	43315000	684871	26018 45 7	26018 45 7	26018 45 7
Moon ..	324000	51229	4523 48 19	4523 48 19	4523 48 19
Mercury	1043208	164946	12868 14 44	60663 28 24	36765 51 34
Venus..	2664637	421316	12879 23 4	304426 10 1	158652 46 32
Mars ..	8146909	1288140	262941 35 0	847254 18 40	555097 56 50
Jupiter..	51375764	8123328	724922 53 2	1631114 5 47	1178018 29 24
Saturn ..	127668255	20186140	215690 0 38	2262997 7 48	1239343 34 8

* The Yojan is about 9.1-11th English miles, according to the *Lilāvati*. But the astronomers reckon the equatorial circumference of the earth about 5059 Yojans: the degree, therefore, becomes equal to 14.1-19th Yojans nearly, which makes the Yojan something less than five miles, taking the degree on the equator to be 69 miles.

No. II.

REMARKS

ON

THE CHINESE ASTRONOMY.

No. II.

REMARKS ON THE CHINESE ASTRONOMY.

In the course of my investigation of the antiquity of the Hindu astronomy, I was induced to take a cursory view of the Chinese, in hopes of finding some analogy between them, and thence be able to draw some conclusion which of the two was the most ancient. In this enquiry, however, I met with nothing that could induce me to believe that any connexion existed between them, at least in ancient times. I found that the Chinese were not only far behind the Hindus in the knowledge of astronomy, but that they were indebted to them, in modern times, for the introduction of some improvements into that science, which they themselves acknowledge. With respect to the Lunar Mansions of the *Chinese*, they differ entirely from those of the *Hindus*, who invariably make theirs to contain 13° 20′ each on the ecliptic; whereas the Chinese have theirs of various extents, from upwards of 30° to a few minutes, and marked by a star at the beginning of each, which makes them totally to differ from the Hindus.

The Arabs were the only people that I knew of beside the Hindus that had Lunar Asterisms; and as they are said to have communicated some of their astronomy to the Chinese about eight or nine hundred years ago, a comparison with their mansions, I thought, might throw some light on the subject, and in this idea I was not mistaken; for, on comparing the Arabian and Chinese Lunar Asterisms together, I found, to my surprise, that not less than thirteen out of the whole number, which consists of 28, were precisely the same, and in the same order, with scarcely a break between them. Here then there appeared sufficient evidence to shew, that there must have been a connexion between them at some former period, and that the one must have borrowed from the other: but the question then was, who borrowed from the other,—was it the Chinese from the Arabians, or the Arabians from the Chinese? If the Chinese were the borrowers, some means might be found of determining the antiquity of the mansions among the Arabs; but if the Arabians were the borrowers, then we must have recourse to the names of the Chinese mansions, to see if they afforded some clue. I mentioned the circumstance to a learned Mahomedan, in hopes of getting some information, and his reply was, that neither the Chinese borrowed from the Arabs, nor the Arabs from the Chinese; but that they both had borrowed from one and the same source, which was from the people of a country to the north of Persia, and to the west or the north-west of China, called Turkistan. He observed, that before the time of Mahomed, the Arabs had no astronomy; that they were then devoid of every kind of science; and that what they possessed since on the subject of astronomy was from the Greeks. To which I replied, that I understood the Mansions of the Moon were alluded to in the Koran; and as the Greeks had no Lunar Mansions in their astronomy, they could not come from them. He said, that the mansions alluded to in the Koran were uncertain; that no one knew what particular star or mansion was meant; and therefore, no inference could be drawn that any of those now in use were alluded to. Here our conversation ended; and as no great light was thrown on the subject, by supposing that the Chinese and Arabs borrowed from one and the same source, instead of one of them from the other, I thought it was best to adopt another course, which was, to examine into the meanings of the names of the Chinese mansions, which might refer us to some of the constellations, in the same manner as the Arabian names of several of their mansions refer to certain parts of the constellations from which they derive their names, by which their antiquity would be, at all events, limited by the period when the constellations themselves were first framed, beyond which they could not be carried, but might otherwise be of a very modern date. With this view, I carefully examined the name of the Chinese constellations, and particularly their mansions, because on the latter only, the antiquity would rest; for of the former many might have been introduced since the first arrival of the missionaries in China, and, perhaps through their assistance; but the latter could not, as they existed before their times. In this search I was not disappointed; for I found that two mansions in Scorpio, Sin, Wei, referred to parts of that constellation: the first, being the name for the Scorpion's heart, is called the *heart* station; the other, signifying tail, is called the *tail* station, thus referring at once to the parts of the constellation to which they respectively belong. We cannot, therefore, on this ground, admit them to possess a greater antiquity than the constellation itself, from certain parts of which they derive their names. Indeed there is strong ground to believe, that they are not older than the third century of the christian era: but before we can enter on this discussion, it is proper to exhibit both the Chinese and the Arabian mansions together, in order to be compared, as in the following Table.

TABLE I.

The Arabian and Chinese Lunar Mansions, exhibiting the first Star of each.

Arabian Names and first Star of each Mansion.		*Chinese Names, and first Star of each Mansion.*	
1 Simakool uazul,	Spica Virg.	1 Keo,	Spica Virginis.
2 Ul Ghufr,	98 Virg.	2 Kang,	98 Do.
3 Uz Zubana,	α Libræ.	3 Te,	α Libræ.
4 Ul Icleel,	π ρ δ β Scorp.	4 Fang,	π ρ δ β Scorpii.
5 Ul Qulb, (Oolugrub,)	α Scorp.	5 Sin,	σ Do.
6 Ush Showlah,	υ Scorp.	6 Wei,	ε Do.
7 Un Nwaim,	γ Sagitt.	7 Ke,	γ Sagittarii.
8 Ul Bulda,	φ Do.	8 Tow,	φ Do.
9 Sad'oodh Dhabih,	β Capric.	9 New,	β Capricorni.
10 Sad'ool Bula,	ε Aquar.	10 Neu,	ε Aquarii.
11 Sad'oos soo-ood,	β Do.	11 Heu,	β Do.
12 Sad'ool ukhbiyuh,	γ Do.	12 Wei,	α Do.
13 Ul Furgh'ool Mooguddim,	β Pegas.	13 Shih,	α Pegasi.
14 Ul Furgh'ool Mooukkbir,	γ Do.	14 Peeh,	γ Do.
15 Ur Risha,	β Andromed.	15 Kwei,	38 Andromedæ.
16 Ush Shurutan,	γ Arietis.	16 Low,	β Arietis.
17 Ul Botyn,	ρ ρ ρ Do.	17 Wei,	35 Br. C. Do.
18 Uth Thuryya,	η Pleiad.	18 Maou,	η Pleiadum.
19 Ud Duburon,	γ Tauri.	19 Peeh,	γ Tauri.
20 Ul Hiqab,	λ Orionis.	20 Chuy,	λ Orionis.
21 Ul Hinah,	γ Geminorum.	21 Tsan,	δ Do.
22 Udh Dhira,	α Geminorum.	22 Tsing,	μ Geminorum.
23 Un Nuttruh,	Præsepe.	23 Kwei,	θ Cancri.
24 Ut Turfuh,	ε Leonis.	24 Lew,	δ Hydræ.
25 Uj Jebhah,	η ζ γ Do.	25 Sing,	α Do.
26 Uz Zoobruh,	72 Br. C. Do.	26 Chang,	υ Do.
27 Us Surfuh,	β Do.	27 Yih,	α Crateris.
28 Ul Awa,	β Virginis.	28 Chin,	γ Corvi.

The first four mansions appear the same with the Chinese and Arabians; the 5th differs in the Chinese beginning the mansions with σ instead of α Scorp. but as including α, they call it the mansion of the heart; in the sixth they also differ, but the 7th, 8th,

9th, 10th, and 11th, are the same; in the 12th and 13th they differ a little, and in the 14th they again agree, but disagree in the 15th, 16th and 17th; in the 18th, 19th, and 20th they also agree, but disagree in all the rest.

Here then we see, that out of the whole no less than 13 agree, and these all within the first twenty. This regular agreement and commencement of both from the same point is so particularly striking, that no one can doubt but that they must have been obtained originally from the same fountain head. That they are not of Chinese origin is certain, because the point of commencement does not agree in any manner with that of the Chinese year, which either begins at the autumnal or vernal equinox; and there is not the slightest doubt but that the year, in the country where the mansions were framed, must have begun either at the vernal or autumnal equinox, but most probably the vernal, as the star in the Lunar Asterism would then be on the meridian at midnight, and the longitude of the star six signs from Aries. If we wish to know when this was the case, we must determine the time from the present longitude of Spica.

In A. D. 1750, the longitude of Spica Virginis was 6ˢ 20° 21′ 18″, so that it has advanced 20° 21′ 18″. Now if we assume the annual precession at 50″ we shall get 1466 years, which, subtracted from 1750, leaves A. D. 284, the time when the autumnal equinox coincided with the star, which we conclude was also the time when the lunar Mansions were framed. Consequently, if this assumption be correct, the Chinese must have received their mansions from some quarter, since the Year A. D. 284, and, in order to prevent detection, gave them names of their own. To this it may be objected, that they have tables of Lunar Mansions of a date many years anterior to A. D. 284; that they are mentioned in their old books; and that celestial observations of very great antiquity are referred to them. To which it may be observed, that all this is fine declamation, full of plausibility, but without the slightest proof. A nation like the Chinese, who are proud and jealous of all others, would naturally use every possible means to conceal the adoption of a set of Lunar Mansions received from another state, to which they would not confess themselves to be under an obligation. They would therefore not only disguise them by new names of their own, but would likewise compute tables of them for different periods of time, long anterior to the time of their reception; and to make the matter still more complete, they would make mention of them in their pretended ancient books, fabricated then, and refer various fictitious celestial observations to them, settled by computation backward, or perhaps by no computation at all, as in the supposed observations on comets, in which they knew they could not be detected. Many of these pretended observations have been found to be false, and the others that were supposed to come near the truth, were no other than the effect of mere computation in modern times. But that which destroys all these supposed ancient references most completely, is the time of the formation of the constellations, which I have placed between the years 756 and 746 B. C., and others carry back as far as 1100 or 1200 years before Christ. If they have imposed on us in this respect, by going beyond the limit here assigned, what security have we that they have not imposed on us in the first instance, in assuming to themselves the invention and formation of the Lunar Asterisms above exhibited, which most certainly never were Chinese, nor were invented before the time I mentioned above, viz. A. D. 284? But whether these mansions were the invention of the people of Turkistan, or of the Arabians, we have not the means to ascertain. The Arabians generally use the Lunar year; but they have also the astronomical solar tropical year, beginning at the vernal equinox, I believe, and therefore there would be no inconsistency in considering them the inventors, until we can discover the real country to which they originally belong. The Chinese themselves admit, in some of their books, that in the year A. D. 104, they received from the West a work containing a catalogue of 2500 stars; but when this book is enquired after, they pretend that it is lost—a circumstance which at best looks very suspicious: that work might contain something that would disclose some of the Chinese impositions, and therefore, it may be presumed, is carefully kept out of sight.

Some time after I had written the above remarks, and was just ready to send them to the press, a book fell into my hands (Kircher's Lingua Egyptiaca), containing the Egyptian Lunar Mansions, which appear to be those we are in search of; for they make the equinoctial colure to cut the star Spica Virginis, which could not be later than A. D. 284, the epoch to which I referred the formation of the Arabian and Chinese Lunar Mansions, as founded on that circumstance.

The following table of Lunar Mansions, with their Egyptian names and explanations, longitudes, &c. are taken from this work.

TABLE II.

	Egyptian Names, &c.	*Long.*	*Arabian Names, &c.*
1	Kuton, Statio Pisces	0^s 0° 0′	Ur Risha, Funis.
2	Pikutorion, Pisces Hori	0 12 24	Ush Shurutan, Fixed mark.
3	Kolion, Stat. Connectens	0 25 0	Ul Botyn, Ventriculus.
4	Orias, Statio Hori	1 9 0	Uth Thuryya, Pleiades.
5	Piorion, Statio Hori Major	0 21 0	Ud Duburan, Hyades.
6	Klusos, Claustrum	2 4 0	Ul Hiqah, Box, Chest.
7	Klaria	0 17 0	Ul Hinah.
8	Pimahi, Cubitus	3 0 0	Udh Dhira, Cubitus.
9	Termelia, Stat. influentia	0 13 0	Un Nuthruh, Watchman, guard.
10	Piautos, Seipsam parturiens	0 26 0	Ul Turfuh, Oculus.
11	Ditehni, Frons	4 9 0	Uj Jebhah, Frons.
12	Pichorion, idem	0 21 0	Uz Zoobruh, Lion's Mane.
13	Asphulia, Statio Amoris	4 4 0	Us Surfuh, Vicissitude.
14	Abukia, Statio Latrantis	0 18 0	Ul Awa, Barking.
15	Choritos, Statio Altitudinis*	6 0 0	Simakool uazul, Altitude α Virginis.
16	Chambalia, Stat. Propitiationis	0 13 0	Ul Ghufr, Pardoning.
17	Prititbi	0 26 0	Uz Zubana, Claws.
18	Stephapi, Corona	7 9 6	Ul Icleel, Corona.
19	Charthian, Cor	0 21 0	Ul Qulb, Cor.
20	Aggia, Sancta	8 4 0	Ush Showlah, Tail.
21	Nimamreh, St. Gratiæ and Jucunditatis	0 17 0	Un Nwaim, Comfort, enjoyment.
22	Polis, Civitas	9 0 0	Ul Bulda, Urbs.
23	Upeutos, Brachium sacrificii	0 13 0	Sa'd ud Dhabih, Fortuna mactantis.
24	Upeuritos, Brachium absorptum	0 26 0	Sad ool Bula, Fortuna deglutientis.
25	Upeuineutes, Beatitudo beatitudinum	10 9 0	Sa'd-oos-soo ood, Fortuna fortunarum.
26	Upeutherian, Brachium absconditum	0 21 0	Sa'd oolukhbiyuh, Fortuna tentoriorum.
27	Artulos, Stat. Prioris germinationis	11 4 0	Ul Furgh ool mooguddim, Evacuatio anterior.
28	Artulosia, Statio Posterior germinatio	0 17 0	Ul Furghool mooukker, Idem posterior.

* By *Altitudinis*, which is also the name in Arabic and Chinese, is meant, that at the commencement of the year, at the vernal equinox at noon, the star Spica is on the meridian at midnight, and of course at its greatest altitude; which points out the age of the mansions to be about 1530 years in A.D. 1823. They could not therefore be known to Ptolemy.

On looking over this table, the first thing that strikes us, as different from any other we have seen, is, that it is divided into four portions or parts of 90° each; the first of which begins with Aries, or the vernal equinoctical point; the second with Cancer, or the summer solstice; the third with Libra, or the Autumnal equinox; and the fourth with Capricorn, or the winter solstice. It also appears, that the extent of each Lunar Mansion is about 13°, and the Spica Virginis is placed in six signs from Aries, which makes the antiquity of the table (now 1823) 1539 years. I have given Kircher's explanations of the Egyptian names of the mansions, in order that they might be compared with the explanations of the Arabic names of the same mansions, which I have taken from the commentary on Ulugh Beigh's Table of the stars; by which it will be seen that they are the same, or nearly so, throughout, and that consequently the Arabs must have borrowed from the Egyptians, and the Chinese from the Arabs; at least I am rather inclined to think so.

As the other parts of the Chinese astronomy afford no criterion for judging of their antiquity, and being in fact generally acknowledged to be modern, we shall now close the subject with the following tables of the Chinese constellations, which I have taken from the Rev. Dr. R. Morrison's Chinese Dictionary, hoping that, as they are not generally known, they may be acceptable to those who take an interest in the study of astronomy.

TABLE III.

Alphabetical Arrangement of the Chinese Constellations and Stars.

1	Chang yuen	k t and small stars, Leo.
2	,, sha	ζ Corvus.
3	,, suh,*	κ υ υ λ λ μ φ φ Hydra.
4	,, chin,	Cor Caroli.
5	,, jin,	α Columba.
6	Chaou yaou,	β Bootes.
7	Chaou,	λ Hercules.
		m Capricorn.
8	Chay foo,	ρ ζ and 2 small stars, Cygnus.
9	,, ke,	π 3324, L 3341, e 3358, Lupus.
10	,, sze,	ν Serpens.
11	Chih new,	α Lyra.
12	Chin suh,*	γ ε Corvus.
13	,, chay,	γ Libra, f Lupus.

14	Choo,	$\alpha\ \phi\ \chi\ \tau\ \nu\ \upsilon$ Auriga, i, g, k, ψ A and another, Centaur.
15	,, she,	χ Draco.
16	Choo,	π Pegasus, α Ara Θ s. name.
17	,, wang,	τ B.C. 1200, 1228, between the horns of Taurus.
18	Chow,	β Serpens, s. n. η Capricornus.
19	,, ting,	$\upsilon\ \omega$ Coma Berenice.
20	Chung tae,	$\lambda\ \mu$ Preceding hind foot of Ursa Major.
	Chuy suh yih,*	λ Orionis.
21	Chung shan,	b $\xi\ \nu$ Hercules' hand.
22	E tseo,	Apus, Bird of Paradise.
23	Fa,	l υ Orion, s. n. l 4862 Capricornus.
24	Fang suh,*	$\beta\ \delta\ \pi\ \rho$ Scorpio.
25	Fe yu,	Piscis Volans.
26	Foo yue,	b 5333, 5345, 5362, 52, Stream of Aquarius.
27	,, chih,	c y near τ, Cetus.
28	,, loo,	$\zeta\ \lambda$ Cassiopeia.
29	,, urh,	Small stars near Aldebaran.
30	,, pih,	γ Hydrus.
31	,, sing,	g small stars near Mizar, tail of Ursa Major.
32	,, shay,	Cluster in hand of Perseus.
33	,, shwo,	γ Telescopium.
34	Fun moo,	$\gamma\ \zeta\ \eta\ \pi\ \tau$ Aquarius.
35	Hae shih,	A Argo?
36	,, shan,	λ and small stars near Crux, and the foot of Centaur.
37	Han,	ζ Ophiucus' knee, s. n. ϕ Capricorn.
38	Hang,	ν 3039 μ 3030 ϕ 3069 Centaur.
39	He chung,	$\kappa\ \theta$ Cygnus.
40	Hea tae,	$\nu\ \xi$ following hind foot of Ursa Major.
41	,, tsae,	θ Draco.
42	Heen yuen,	Regulus, A 2232 $\gamma\ \epsilon\ \lambda\ \eta\ \rho\ \nu\ o\ \zeta\ \kappa$ Leo.
43	Heu suh,*	β Aquarius.
44	,, leang,	κ Aquarius.
45	Heuen ko,	γ Bootes.
46	Hin chin,	2629 Coma Berenice, near E, Leo.
47	Ho neaou,	Phœnix.
48	,, këen,	γ Hercules.
49	,, chung,	β Hercules.
50	,, koo,	Altair $\beta\ \gamma$ Aquila.
51	Ho,	Grus.
52	Hoo fun,	t 2470, near δ Leo.
53	,, she,	$\delta\ \epsilon\ \kappa\ \eta$ Canis Major, and $\delta\ \omega$ Argo.
54	,, kwa,	a $\beta\ \gamma\ \delta\ \zeta$ Delphini.
55	How kung,	b 3162 Ursa Minor.
56	How,	a Ophiucus Ras Alhaque.
57	Huh,	$\iota\ \kappa$ Ophiucus, k i Hercules, near ditto.
58	Hwa kae,	4 stars bet. Cassiop. and Camelop. uncertain.

59 Hwan chay, e i and two small stars near chin of Ophiucus.
60 Jih, κ Pegasus μ Cygnus, s. n. κ λ Libra.
61 Jin uh, e f g Pegasus, near fore foot.
62 Kae uh, o Aquarius.
63 ,, yang, ζ Mizar Ursa Major.
64 Kang che, 4 small stars near Arcturus.
65 ,, suh,* ι κ λ μ ρ Virgo.
66 Kang, p 3947, Sagittarius.
67 Kang ho, ρ δ Bootes.
68 Ke suh,* First γ 4053, δ Sagittarius.
69 ,, wan, κ Centaurus, β Lupus.
70 Ke chin }
70 tseang keun, } ρ near the Rump of Lupus.
71 Keaow pih, θ Dorado.
72 Keen sing, ν ξ o p s Sagittarius' head.
73 ,, pe, υ Scorpio.
74 Keih, θ Aquarius.
75 Keo suh,* Spica, ζ Virgo.
76 Keue kew, k m Monoceros.
77 Keuen she, ν Perseus.
78 Kean nan mun, χ φ Andromeda.
79 Kung tsing, ι κ λ ν Lepus' Ears.
80 Keun she, β Canis Major.
81 Kew hiang, ρ and small stars, Virgo.
82 ,, ho, μ Hercules' arm.
83 ,, yew, μ ω b Eridanus, and stars in Sceptrum Brandenburg.
84 ,, chow choo yih A o d c ξ r &c. Eridanus.
85 Kih sing, New star in Cassiopeia.
86 Kin yu, ε Dorado, probably the whole.
87 Ko taou, υ ξ o π Cassiopeia.
88 Koo low, γ τ Centaurus.
89 Kow ching, ζ Ursa Minor.
90 Kow, ψ 4322, two χ 4364, 4365 Sagittarius.
91 ,, kwo, μ b a c Sagittarius.
92 ,, ling, Two ω Scorpio.
93 Kuh, μ Capricornus.
94 Kung tseo, Pavo.
95 Kwan, λ μ χ, three φ ω ψ Cancer.
96 Kwan Soo, Corona Borealis.
97 Kwei suh,* δ ε ζ η μ ν π and Mirack, β Andromeda υ φ χ, and two ψ Pisces.
98 ,, ,, * λ δ η θ Cancer.
99 Lang wei, a b c d e f, Coma Berenice.
100 Lang tseang, p Do.
101 Laou jin, Canopus.
102 Le kung, λ μ, τ υ, and n o Pegasus.
103 ,, shih, χ φ ψ Taurus.
104 Leang, δ Ophiucus, Yed.

105	Lee tsze,	λ Ophiucus.
106	Leen taou,	η θ Lyra.
107	Lew suh,*	δ ε ζ η θ ρ σ ω Hydra.
108	Ling tae,	χ c d Leo.
109	Lo suh,*	υ Capricornus.
110	Low suh,*	a β γ Aries' head, &c.
111	Luh kea,	Stars bet. Tarandus and Camelop.?
112	Luy teen,	ζ Pegasus.
113	,, peih chin,	ε κ γ δ Capric. ι σ λ φ Aquar. p s q 5476 Pisces.
114	Ma we,	δ Centaurus.
115	Ma fuh,	β Do. W. foot.
116	Maou suh,*	Pleiades.
117	Meih fung,	Musca Australis.
118	Ming tang,	τ υ φ e r, Leo.
119	Nan ho,	α, β η Canis Minor.
120	Nan chuen,	θ &c. Robur Caroli?
121	Nan mun,	a Centaurus, E. foot.
122	Nan hae,	ξ and 2927 Serpens.
123	Neaou hwuy,	δ Toucan, perhaps the whole.
124	Neu tsang,	e π ρ Hercules.
125	,, she,	ψ Draco.
126	,, suh,*	ε μ ν Aquarius.
127	New suh,*	α β and Neb. 323 324 Capric. and Neb. 322 Sagittarius.
128	Nuy keae,	τ i and small stars between Eye and Nose Ursa Major.
129	,, ping,	ν ο π ξ Leo Minor? if not Virgo.
130	,, ping,	ν ο π-ξ Virgo.
131	Pa,	ε Serpens.
132	Pa keuh,	δ ξ h k i head of Auriga D D f near Cassiopeia.
133	Pae kwa,	ε Delphinus,
134	,, kew,	λ γ Grus.
135	Pae,	Corona Australis.
136	Peih suh,*	Algenib γ Pegasus.
137	,, leih,	β γ θ ι ω Pisces.
138	,, suh,*	Hyades Aldebaran γ δ ε λ ο.
139	Pih too,	C B P Q Cerberus' head.
140	,, ho,	ρ σ Gemini.
141	,, lo sze mu,	Fomalhaut.
142	Ping taou,	t θ Virgo.
143	,, sing,	ν Hydra, h Centaurus.
144	,, sing,	ε μ Lepus.
145	Po sing,	α Indus?
146	San sze,	ρ σ σ near Ear of Ursa Major.
147	San kung,	Three small stars bet. γ δ η Virgo, s. n. to 3 stars near Asterion's head.
148	,, keo hing,	Southern triangle.

149 Se han, ε ζ η θ ξ e Libra.
150 Seang, Small stars bet. δ and ε Ursa Major.
151 Seaou tow, Cameleon.
152 Seu, θ Serpens.
153 Shang ching, A Camelopardalis.
154 Shang wei, L Do. s. n. κ Cepheus.
155 Shang tae, ι κ. Fore foot of Ursa Major.
156 „ tseang, σ Leo, s. n. υ Coma Berenice.
157 „ seang, δ Leo, s. n. γ Virgo.
158 „ foo, λ Draco.
159 „ peeh, ζ Draco.
160 „ shoo, A 3687.
161 Shaou wei, γ Cepheus, s. n. C. Camelopard.
162 „ ching, n Tarandus.
163 „ foo, d Ear of Ursa Major.
164 „ wei, m Leo, and m p r Leo Minor.
165 „ foo, χ Ursa Major.
166 „ tsae, η 2348 Draco.
167 „ peih, ψ 2348 Draco.
168 Shay fuh, Small stars between Hydrus and Toucan.
169 „ show, ε ζ Hydrus.
170 „ we, β Octans.
171 She, Small stars near leg of Columba.
172 „ low, μ Ophiucus.
173 Shih tsze kea, Crux.
174 „ suh,* a Pegasus, Markab.
175 Shin kung, ζ 3739 and 3745, Scorpion's tail.
176 Shuh, α λ Serpens.
177 Shwny foo, ν ξ Orion's hand.
178 „ wei, ζ θ ο p Canis Minor.
179 „ low, a Eridanus, Achernar.
180 Sin suh,* Antares σ τ and two c and 7 Scorpio.
181 Sing suh,* Alphard, a Hydra and small stars near.
182 Sun, θ κ Columba.
183 Sung, η Ophiucus.
184 Sze kwae, H Taurus χ χ Orion.
185 Sze wei, α β Equuleus.
186 „ fe, γ δ Do.
187 „ fuh, b f g i Monoceros.
188 „ foo, N and small stars near head of Camelopard.
189 Ta ling, τ Perseus.
190 „ tsun, δ Gemini.
191 „ keo, Arcturus.
192 Tae tsun, ψ Ursa Major.
193 „ yang show χ Ursa Major.
194 „ yih, i Small stars near α Draco.
195 „ tsze, γ Ursa Minor, s. n. E, Leo.
196 Tang mun, b c c Centaur.
197 „ shay, π Cygnus and stars near.

198	Te,	Kochab, Ursa Major.
199	,, tso,	a Hercules, Ras Algethi.
200	,, suh,*	$\alpha\ \beta\ \gamma\ \delta\ \iota\ \mu\ \nu$ Libra.
201	Teen choo,	α Ursa Major, Dubhe.
202	,, choo,	$\delta\ \epsilon\ \pi\ \rho\ \sigma$ Draco.
203	,, chuen,	$\gamma\ \eta$ Perseus.
204	,, fow,	θ Antinous.
205	,, fuh,	d o Scorpio.
206	,, han,	Via Lactea.
207	,, hwan,	Four ϕ Cetus.
208	,, hwang,	$\mu\ \rho\ \sigma$ near λ Auriga.
209	,, hwang ta te,	Pole Star.
210	,, joo,	Second α, ω Serpens.
211	,, kaou,	ι Taurus.
212	,, ke,	λ Ursa Major.
213	,, ke,	e f Sagittarius.
214	,, ke,	Small stars near θ Hercules, s. n. ψ Argo.
215	,, keae,	$\kappa\ \upsilon$ Taurus' Ear.
216	,, kew,	$\theta\ \rho\ \sigma$ Andromeda's Arm.
217	,, keang,	$\beta\ \theta\ \rho$, e Ophiucus, and a b Sagittarius.
218	,, keun,	α Menkar, $\gamma\ \delta\ \lambda\ \mu\ \nu$ two ξ Cetus.
219	,, kow,	1971 Argo.
220	,, kwan,	1217, 1192 ζ Taurus.
221	,, kwan,	δ Ursa Major, 97, if not ρ.
222	,, lang,	Sirius, α Canis Major.
223	Tëen laou,	ω, and stars near, in Ursa Major.
224	,, le,	Four stars within the ▭ of Ursa Major.
225	,, lin,	ξ e f G o Taurus.
226	,, luy ching,	ξ Aquarius, λ Capricornus and other small stars.
227	,, mun,	2395, p r Leo.
228	,, o,	e Aries.
229	Teen pëen,	$\iota\ \lambda$, h g. Foot of Antinous, and stars in Scutum Sobieski.
230	,, seang,	q Sextans.
231	,, seuen,	β Ursa Major.
232	,, shay,	η Argo.
233	,, ta tseang keun,	Triangle, includes λ Andromeda and other small stars.
234	Tëen tëen,	$\sigma\ \tau$ Virgo.
235	,, tsan,	$o\ \zeta$ Perseus' foot.
236	,, tsang,	$\iota\ \eta\ \theta\ \tau$ Cetus.
237	,, tsang,	$\iota\ \theta\ \kappa$ hand of Bootes.
238	,, tsëe,	$\pi\ \rho$ b h c small stars near Hyades.
239	,, tsin,	γ Cygnus.
240	,, wang,	$\beta\ \rho\ \zeta$ Piscis Notius.
241	,, yih,	χ Draco.
242	,, yin,	$\delta\ \zeta$ Aries.

243	Teen yu,	Small stars in Fornax Chemica.
244	„ yuen,	ι κ χ φ Eridanus.
245	„ yuen,	γ δ ε ζ η τ, E l m t, Eridanus.
246	„ yueñ,	β a H K Sagittarius.
247	„ tsëen,	η θ ι μ Piscis Notius.
248	To ming,	d Piscis.
249	To kung se,	d Pegasus.
250	Too sze kung,	β Cetus.
251	Too sze,	D F. Cerberus' head.
252	Tow,	ω, h g n o Hercules, near hand and club.
253	Tow suh,	ξ λ σ τ φ Sagittarius.
254	Tsan ke,	o o ζ, &c. Lion's Skin, Orion.
255	Tsan suh,	Betelguese, Bellatrix, Rigel, γ δ ε ζ κ Orion.
256	Tsaou foo,	δ ε ζ Cepheus.
257	Tse,	H Hercules, near Cerberus.
258	Tseih she,	Caput Medusæ,
259	„ shway,	λ μ Perseus.
260	„ sin,	χ Gemini, μ Cancer.
261	„ she ke,	Presepe, in Cancer.
262	Tseih,	γ and another star, Lupus.
263	Tseen tae,	β δ ι Lyra.
264	Tsew ke,	ξ ψ ω Leo 2083, κ ξ Cancer.
265	Tsih,	γ Cassiopeia.
266	Tseih kung,	δ μ ν ψ χ χ club of Bootes.
267	Tsin heen,	ψ χ, g k Virgo.
268	Tsin,	κ q Hercules.
	„	b Capricornus.
269	Tsing kew,	β ξ v Hydra.
270	Tsing suh,	γ ε ζ λ μ ν Gemini.
271	Tso kang,	ε Aries.
272	Tso cheh fa,	η Virgo.
273	„ she te,	ξ ο π ξ Bootes.
274	„ choo,	ι Draco.
275	„ kea,	δ Algorab, β η Corvus.
276	„ ke,	ξ Aquila.
277	„ ke,	y, and stars near hand of Auriga.
278	Tsoo,	ε Ophiucus.
	„	A Capricorn.
279	„ kaou,	ε ρ σ Cetus.
280	Tsow,	δ Serpens.
	„	θ Capricorn.
281	Tsung kwan,	2567 Leo.
	„ „	χ φ φ Lupus.
282	„ ching,	β γ Ophiucus.
283	„ jin,	k n o p q Taurus, Poniatowski.
284	„ sing,	K M N O Hercules.
285	Tsze tseang,	l Leo.
	„ „	ε Virgo.

286	Tsze seang,	θ Leo.
	,, ,,	δ Virgo.
287	,,	λ Columba.
288	,,	a β γ δ Lupus.
289	,, suh,	λ Orion.
290	Tung han,	φ χ ψ ω Ophiucus' foot.
291	,, hae,	η ζ Serpens.
292	Wae ping,	a δ ε ζ μ ν ξ Pisces, Fish band.
293	,, choo,	q r Monoceros' tail.
294	Wan chang,	φ θ υ Ursa Major, fore leg.
295	,, lang,	a β η κ Cassiopeia.
296	Wei suh,	Musca.
297	,,	a Hercules.
	,,	χ Capricorn.
298	,,	d Telescopium.
299	,, suh,	a Aquarius, ε Pegasus.
300	,, suh,	ε μ Scorpio,
301	Woo chay,	Capella β θ κ Auriga, and β Taurus.
302	,, choo how	θ ι υ τ φ Gemini.
303	,, te tso,	β Leo, and four small stars near.
304	,, tsze,	a Ursa Minor.
305	,, yue,	ε ζ Aquila.
306	Yang mun,	a Lupus.
307	Yaou kwang,	η Ursa Major.
308	Ye ke,	ο π, and small stars, Canis Major.
309	Yen,	ν ζ Ophiucus.
	,,	ζ Capricorn.
310	Yew kang,	η ο ρ π and star near, Pisces.
311	,, chih fa,	β Virgo.
312	,, she te,	η υ τ Bootes.
313	,, choo,	a Draco.
314	,, pea,	a Corvus.
315	,, ke,	δ η ι κ Antinous.
316	Yin tih,	Q Camelopardus.
317	Yu neu,	π Leo.
318	Yih suh,	a Crater.
319	Yu,	y Ophiucus.
320	Ye lin keun,	δ τ χ ψ ψ ψ Aquarius.
321	Yue,	A Taurus, between Pleiades and Hyades.
322	,,	η Gemini.
323	,,	ψ Capricorn.
324	Yun yu,	κ λ Pisces.
325	Yuh tsing,	β λ ψ Eridanus, τ Orion.
326	,, kang,	ε Ursa Major.
327	Ye chay,	c Virgo.

NAMES OF THE PLANETS.

King sing,	Venus.	Shwuy sing,	Mercury.
Muh sing,	Jupiter.	Ho sing,	Mars.
Too sing,	Saturn.		

TABLE IV.

CHINESE NAMES,

***Right Ascensions, Declinations, Longitudes, and Latitudes of Ninety-two Stars, for the Year A. D.* 1683.**

	Chinese Names.	*R. A.*	*Declin.*	*Long.*	*Lat.*	*European Names.*
1	Tung hae	9s 1°18′	2°51′S.	9s 1°23′	20°38′N.	η Serpentis.
2	Chih neu yih	9 6 18	38 22 N.	9 10 27	61 48 —	a Lyræ.
3	Tow suh yih........	9 6 23	27 12 S.	9 5 50	3 50 S.	φ Sagittarii. (1st Mans.)
4	Yu	9 10 7	3 53 N.	9 11 20	26 59 N.	θ Serpentis.
5	Teen yun	9 12 22	5 15 S.	9 12 56	17 41	λ Antinoi.
6	Yew ke san	9 17 21	2 36 N.	9 19 11	24 56	δ Aquilæ.
7	Yew ke luh	9 17 58	7 37	9 20 27	14 28	κ Antinoi.
8	Yew ke woo........	9 20 6	1 45 S.	9 21 28	20 15	ι Antinoi.
9	Ho koo san	9 22 51	9 54 N.	9 26 36	31 18	γ Aquilæ.
10	Ho koo urh	9 23 50	8 7	9 27 19	29 22	a Aquilæ.
11	Yew ke tung tseih ..	9 24 3	8 15	9 26 0	21 38	η Antinoi.
12	Ho koo yih	9 24 57	5 45	9 28 3	26 50	β Aquilæ.
13	Yew ke tung pa	9 28 46	1 41 S.	10 0 32	18 48	θ Antinoi.
14	New suh yih........	10 0 46	15 42	9 29 37	4 41	β Capricorn. (2)
15	Teen tsin yih	10 2 46	39 18 N.	10 20 35	57 10	γ Cygni.
16	Neu suh yih........	10 7 41	10 33 S.	10 7 23	8 10	ε Aquarii. (3)
17	Heu suh yih........	10 18 44	6 52	10 19 1	8 42	β Aquarii. (4)
18	Wei suh san........	10 22 12	8 28 N.	10 27 32	22 8	ε Pegasi.
19	Wei suh yih........	10 27 26	1 48 S.	10 29 0	10 42	a Aquarii. (5)
20	Fun moo sze	11 1 21	2 55	11 2 20	8 18	γ Aquarii.
21	Luy teen yih	11 9 7	8 11 N	11 13 54	15 44	ζ Pegasi.
22	Peh lo sze mun......	11 9 56	31 13 S.	10 29 22	21 0 S.	Fomalhaut. a Pis. Aus.
23	Shih suh yih........	11 12 17	13 33 N.	11 19 7	19 26 N.	a Pegasi. (6)
24	Peih suh yih........	11 29 18	13 26	0 4 48	12 35	γ Pegasi. (7)
25	Teen tsang yih	0 0 53	10 33 S.	11 26 33	10 1 S.	ι Ceti.
26	Too sze kung	0 6 54	19 44	11 28 6	20 47	β Ceti.
27	Kwei suh yih	0 10 10	21 47 N.	0 17 54	15 58 N.	η Andromedæ. (8)
28	Teen tsang san......	0 17 8	9 49 S.	0 11 53	15 47 S.	θ Ceti.
29	Low suh yih........	0 24 18	19 15 N.	0 29 33	8 29 N.	β Arietis. (9)
30	Wei shen tseih......	0 26 27	1 14 N.	0 24 58	9 5 S.	a Piscium.
31	Teen kwan kew	1 5 54	1 5 S.	1 3 12	14 32 —	δ Ceti.
32	Wei suh yih........	1 6 17	26 20 N.	1 12 33	11 16 N.	a Muscæ. (10)
33	Teen kwan pa	1 6 47	1 52	1 5 4	12 3 S.	γ Ceti.
34	Teen yuen luh	1 10 16	10 11 S.	1 4 20	24 34	η Eridani.
35	Teen kwan yih......	1 11 30	2 50 N.	1 9 57	12 37	a Ceti, Menkar.
36	Teen yuen woo......	1 15 10	10 2 S.	1 9 26	25 57	ζ Eridani.
37	Teen yuen sze	1 19 36	10 32 —	1 13 35	27 47	ε Ditto.
38	Maou suh yih	1 21 20	23 3 N.	1 24 48	4 10 N.	ε Pleiadum. (11)

	Chinese Names.	*R. A.*	*Declin.*	*Long.*	*Lat.*	*European Names.*
39	Teen yuen san	1ˢ21°55′	10°54′ S.	1ˢ16°17′	28°47′ S.	δ Eridani.
40	Peih suh yih........	2 2 34	18 26 N.	2 4 3	2 37	ε Tauri, γ Tauri. (12)
41	Yuh tsing san	2 13 5	5 33 S.	2 10 52	27 55 S.	β Eridani.
42	Woo chay urh	2 13 21	45 38 N.	2 17 26	22 52 N.	a Auriga, Capella.
43	Tsan suh tseih	2 14 53	8 38 S.	2 12 27	31 12 S.	β Orionis, Rigel.
44	Tsan suh woo	2 17 4	6 0 N.	2 16 33	16 53	η Orionis, Bellatrix.
45	Tsan suh shih kew....	2 17 13	2 45 S.	2 15 48	25 37	γ Do.
46	Tsan suh yih	2 19 2	0 36	2 18 1	23 38	δ Do. (13)
47	Chuy suh yih	2 19 31	9 40 N.	2 19 22	13 26	λ Do. (14)
48	Fa urh	2 19 58	5 39 S.	2 18 35	28 45	i Do.
49	Fa san	2 20 3	6 11 —	2 18 38	29 17	υ Do.
50	Tsan suh urh	2 20 4	1 27 —	2 19 4	24 34	ε Do.
51	Tsan suh san	2 21 13	2 10 —	2 20 17	25 22	z Do.
52	Tsan suh luh........	2 23 13	9 50 —	2 22 0	33 8	κ Do.
53	Tsan suh sze	2 24 36	7 17 N.	2 24 22	16 6	a Do. Betelguese.
54	Tsing suh yih	3 1 0	22 36 —	3 0 55	0 53	μ Geminorum. (15)
55	Teen lang..........	3 7 50	16 16 S.	3 9 46	39 30	a Canis Major, Sirius.
56	Nan ho urh	3 17 32	8 51 N.	3 17 50	13 34	β Canis Minor.
57	Nan ho san	3 20 44	6 0	3 21 29	15 57	υ Do. Procyon.
58	Pih ho san	3 21 28	28 46	3 18 51	6 40 N.	β Geminorum, Pollux.
59	Wae choo yih	4 1 20	3 10 S.	4 4 20	23 0 S.	q Monoceri B v?
60	Kwei suh yih	4 3 24	19 8 N.	4 1 20	0 48	θ Cancri. (16)
61	Lew suh yih........	4 5 15	6 45	4 5 56	12 27	δ Hydræ. (17.)
62	Sing suh yih........	4 18 3	7 19	4 22 56	22 24	a Hydr. Alphard. (18)
63	Chang suh yih	4 24 3	13 29	5 1 19	26 12	v Hydra? (19)
64	Kan yuen shih sze ..	4 27 51	12 30 N.	4 25 25	0 27 N.	a Leonis, Regulus.
65	Teen seang	4 29 20	4 34 S.	5 3 7	16 0 S.	
66	Yih suh yih	5 11 9	16 37 —	5 19 23	22 41 —	α Crateris. (20)
67	Nuy ping yih	5 22 14	10 1 N.	5 18 54	6 7 N.	ξ Virginis.
68	Woo te tso	5 23 13	16 21	5 17 13	12 18	β Leonis, Denebola.
69	Yew chih fa........	5 23 35	3 34	5 22 42	0 43	β Virginis.
70	Chin suh yih........	5 29 58	15 44 S.	6 6 23	14 25 S.	γ Corvi, Algorab. (21)
71	Tung shang seang ..	6 6 25	0 18 N.	6 5 46	2 50 N.	γ Virginis.
72	Tung tsze tseang	6 9 56	5 8	6 7 5	8 40	δ Do.
73	Keo suh yih........	6 17 10	9 27 S.	6 19 26	1 59 S.	a Do. Spica. (22)
74	Keo suh urh........	6 19 39	1 5 N.	6 17 43	8 42 N.	z Do.
75	Kang suh yih	6 29 0	8 44 S.	7 0 3	2 58 —	κ Do. (98) (23)
76	Ta keo	7 0 22	20 56 N.	6 19 50	21 3 —	a Bootes, Arcturus.
77	Te suh yih	7 8 24	14 59 S.	7 10 41	0 26	α Libræ. (24)
78	Te suh sze	7 15 2	8 9	7 14 58	8 35	β Do.
79	Kwan seo yih	7 20 20	27 50 N	7 7 49	14 23	a Coronæ Bor.
80	Suh	7 22 14	7 30	7 17 40	25 36	a Serpentis.
81	Pa................	7 23 50	5 30	7 19 57	24 6	ε Do.
82	Fang suh yih	7 24 57	25 7 S.	7 28 31	5 23 S.	π Scorpio. (25)
83	Leang	7 29 30	2 50	7 27 55	17 10 N.	δ Ophiuci, yed.
84	Tsoo..............	8 0 27	3 51	7 29 7	16 31	ε Ophiuci.
85	Sin suh yih	8 0 29	24 43	8 3 21	3 55 S.	σ Scorpio. (26)
86	Han	8 4 58	9 50	8 4 49	11 30 N.	z Ophiuci.
87	Wei suh yih........	8 6 42	36 57	8 10 54	15 0 S.	ε Scorpio. (27)
88	Te tso	8 15 50	14 45 N.	8 12 40	37 23 N.	a Hercules.
89	Tsung ching yih	8 21 58	4 47	8 20 55	28 1	β Ophiuci.
90	Tsung ching urh	8 23 3	2 53	8 22 15	26 11	γ Do.
91	Ke suh yih	8 26 21	30 23	8 26 50	6 56 S.	γ Sagittarii. (28)
92	Tsung jin nan shih woo	8 26 0	3 29	8 25′45	19 57 N.	z Serpentis.

No. III.

TRANSLATIONS OF CERTAIN HIEROGLYPHICS

WHICH HITHERTO HAVE BEEN CALLED

(THOUGH ERRONEOUSLY)

THE ZODIACS OF DENDERA

IN EGYPT.

No. III.

Translations of certain Hieroglyphics, which hitherto have been called (though erroneously) the Zodiacs of Dendera in Egypt.

In consequence of the extraordinary high antiquity assigned by some of the French writers to those hieroglyphic sculptures called Zodiacs, found in the temple of Dendera, or Tentyra, in Egypt, I was induced some years ago to examine them minutely, and found that, so far from their being Zodiacs, as represented, or called, they were nothing more nor less than the Roman Calendar for the year 708 of Rome, translated into hieroglyphics. This circumstance gave me hopes that re-translating them would be useful in developing the Egyptian method of hieroglyphics, in representing things by their supposed images, particularly such articles as are generally inserted in calendars, which might ultimately lead to a more extensive knowledge of the subject.

The circumstance which appears to have deceived the French writers into an idea of their being zodiacs, and of an antiquity of 15,000 years or more, is simply this. They found that they contained figures of the constellations, that is, outlines without stars; and that some of these figures were again repeated or represented at about the distance of six signs from the original ones of the same name. The former figures they took for the constellations; but the latter they assumed to be the signs, which therefore would require a space of time equal to 15,000 years to bring them into the positions they stand in at present.

Thus they found that the constellation Aquarius, or the figure representing either it or the sign Aquarius, was in its proper place between Capricorn and Pisces. They found also another figure of it under Leo in the circular calendar, and another figure in a boat in the Calendar of the Portico, and from thence drew their conclusion of the extraordinary antiquity of these sculptured zodiacs as they conceived them to be.

It is well known to astronomers, that the constellations appear to rise and set differently at different times of the year. Thus, when the sun is in the same part of the heavens with a zodiacal constellation, that constellation will then appear to rise or set with the sun; and the time of such rising or setting would be recorded in the calendar, and all the risings or settings of the constellations with the sun, would follow each other in regular succession. But when the sun gets round to the opposite part of the heavens, then the same constellation would appear to rise at sunset, and to set at sunrise: the time of the year of such observation being inserted in the calendar, it will be found of course to differ about six months from the former. There are other risings and settings of the constellations which it is not necessary to mention, because their effect is to be considered in the same way. Now suppose this calendar is to be translated into hieroglyphics, with all the different risings or settings of the constellations sculptured on stone, according to the different times of the year at which they occurred, such translation would be made by putting the figure of the constellation in those very places where the name of the same is in writing: consequently the figures of the same constellation would appear in different situations, and at six signs distant from the original. Thus the situation of Aquarius is between Capricorn and Pisces: but according to the Roman calendar, Aquarius sets on the 25th of July, about six signs distant from the situation of the original. Now this is the very figure given in the supposed Zodiac of Dendera, even with the very date attached to it, (see No. 56, in the Calendar of the Portico, Pl. vii;) and all that was done in the translation into hieroglyphics, was to substitute the figure of the constellation in the room of the name, and attach the date to it; which is represented by the figures 5. 5. 1. 1+8. 3. 1. 1=25th of July. There is another date which refers to the 13th of August, the time for which Aquarius is marked in the circular calendar as entirely setting, (see No. 50, Pl. viii.) He is there placed near the figure of Diana with her bow, whose day in the Roman calendar is the 13th of August, and accordingly so marked in the hieroglyphic circular calendar, by the figures 5 and 8=13, underneath, with another figure of Diana with the crescent on her head, No. 49.

Having thus far explained the cause of the deception, in respect to the supposed extraordinary antiquity of those sculptured Roman calendars, I shall now proceed to give a general outline of their contents, which will render them easier to be understood afterwards.

FIRST. *The Date.*

Both the calendars, that is, of the portico, Pl. vii. and circular one of the interior of the temple, Pl. viii. begin with the date 708 (of Rome), at the instant of midnight, and conjunction of the sun, and moon at Rome.

SECOND. *The Phases of the Moon.*

The new and full moons and quarters are occasionally marked throughout that year by a variety of symbols denoting the moon; such as,

1st.—An oval or circle, generally placed on the head of some figure. See Plate vii. No. 31, 39, 48, 55, 60, &c. Plate viii. No. 12, 19, 27, 57, 58, 62, &c.

2nd.—An oval or circle, including within its disc some other figures, in allusion to the spots on the moon. See Plate vii. No. 22, 77. Plate viii. No. 7, 17, 23.

3rd. By the figure of an eye, whether included within a circle or not. See Plate viii. under No 1; also No. 23, and under 73.

4th. By the figure of an animal like a sheep, sometimes with a circle or oval on its head. See Plate viii. No. 12, 60, 86.

5th. By a bird, particularly the Ibis, whether united to any other of the symbols or not. See Plate vii. No. 77, 91. Plate viii. No. 60, 78.

6th. By the figure of a man, with the head and beak of an Ibis. See Plate viii. No. 1. Plate vii. No. 53.

7th. By the figure of a fish? See Plate viii. No. 8, 9, and 16.

There may be other symbols of the moon, which the reader may discover.

THIRD. *The Sun's entrance into the Signs.*

The day on which the sun enters the sign is sometimes marked by a figure of a man with a hawk's head, as a symbol of the sun; sometimes by a female figure, and sometimes by other figures intended to represent the sun. Besides these, there is also a mark to express the sun's ingress into the sign, (see Plate ix.) All these figures have in general the day of the month marked in tablets close to them, except the last, which is to be found in the middle column only, but with the day marked close to it. The sun enters the sign twice in every month, on two different days. Thus the sun enters the sign Capricorn on the 18th December, but the winter solstice is marked on the 25th, making a difference of seven days. The cause of this apparent inconsistency I shall endeavour to explain hereafter. See the Roman calendar, as also Columella, at the four cardinal points of the year, or rather the months of December, March, June and September.

FOURTH. *The Festivals and Agonalias.*

The festivals and agonalias, (or sacrifices,) are generally marked by such figures as seem best to convey an idea of the thing intended; and the days of the month on which the same occur are always marked. See Plate vii. No. 7, 8, 33, and 97; and Plate viii. No. 4, 5, 24, 36, and 83, for the agonalias or sacrifices. The number of days the festival lasts is generally marked by as many stars.

FIFTH. *The Rising and Setting of the Constellations.*

The rising and setting of the constellations are simply marked by their figures, sometimes with dates, and sometimes not, the situation in most cases being sufficient to point out the time nearly, as well as the kind of rising or setting. The figures of the zodiacal constellations, as they stand in order, most probably are meant to represent the signs only, there being no regard paid to proportion, the manner or positions in which they stand, or their distances from each other, or from other known figures near them.

SIXTH. *The Seasons of the Year.*

The seasons are generally marked with the figures of Anubises, or figures of men or animals of any description, with dogs' heads or faces, or dogs' feet. These figures are of a great variety of shapes, but invariably marked with something canine, which makes them easily known at first sight. Some are female, but mostly male. Whether the difference in their shapes depends on any particular time of the year, I have not been able to ascertain with sufficient precision. When they represent the time of harvest, they are figured with an instrument to reap or cut down the corn. See Plate viii. No. 59, 61, 75. The seasons appear to vary, being sometimes earlier, sometimes later by two months, which appear to be all referred to, not only in the Roman calendar, but also in the hieroglyphic translations of it, which therefore obliges me to give here a table of the different seasons for the sake of reference.

Table of the different Seasons.

	A	B	C	D
Spring begins	11 January,	9 February,*	23 February,	12 March.
Middle	25 February,	25 March,*	9 April,	25 April.
Summer begins	11 April,	13 May,*	25 May,	13 June.*
Middle	24 May,	24 June,*	9 July,	25 July.
Autumn begins	11 July,	11 August,*	24 August,	12 September.
Middle	24 August,	24 September,*	9 October,	25 October.
Winter begins	11 October,*	12 November,	24 November,	12 December.
Middle	24 November,	25 December,*	10 January,	25 January.

The * refers to the Roman calendar, where the dates here arranged in order will be found. Those not marked are known by considering that a whole season is, on a medium, 91¼ days, and half a season 45 or 46 days. The letters A, B, C, and D, at the heads of the columns, serve for the purpose of referring to the particular season intended in the following translation.

SEVENTH. *Birds of Passage.*

These, as being connected with the seasons, are also represented in the hieroglyphic calendars by the figures of the birds being given, and the times of the year marked. But unfortunately, as they are not all named in the Roman calendar brought down to our time, we cannot say with certainty what birds are intended by the different figures. The swallow, I believe, is the only one marked in the Roman calendar, and is set down as appearing on the 23rd of February; and so we find it marked in the hieroglyphic calendar plate vii. under No. 24. In the Roman calendar the departure of the swallows is marked the 15th of September, and they continue to disappear for one month, or until the 15th of October. These circumstances are also marked in the circular calendar, plate viii. thus. On the 15th September is the figure of a bird, No. 63, with expanded wings ready to depart; and in October, nearly under the figure No. 72, is that of a bird, No. 71, with one wing expanded; and just before it, the figure of a man sitting: the meaning of which, as will be hereafter shown, is, end, termination, cessation, pause, stop, &c. thereby indicating the last appearance of the swallows.

It is needless to say more on the subject of birds of passage, as we have not the means within our reach of specifying them in a more particular manner. It is proper to notice here, that there are other figures of birds in the hieroglyphic calendars that have no connexion whatever with the birds of passage, and therefore must be carefully distinguished, to avoid falling into error. These are, the eagle or hawk, a symbol of the sun: birds representing the number twenty, which are in general connected with other numbers; and figures of birds symbolic of time, of the moon, and of the month.

EIGHTH. *Numbers and Dates.*

These form the most essential parts of the calendars, because it is through their means we are in a great measure enabled to develope the meanings of all the principal figures, which otherwise would be unattainable, or at most could only be conjectured. It is to be regretted that the whole of the tablets are not legible, the numbers in many of them being worn out by time. However, so far as they are clear and distinct, I have given them in Plate ix. as also the compound numbers, with their explanations.

It appears that the Egyptians had two modes of dating. When the number exceeded a hundred, they used running figures, as we do at present, reading them from left to right: thus, in the date of the years at the heads of the calendars, we see 708, seven hundred and eight: but in small numbers under a hundred, as in the days of the months, they made use of a different method. In this case they employed any number of figures, all the same or different, to make up by *addition* the number required. Thus, suppose the date to be represented was twenty-four, they would not write 24, but take those numbers that would, by their *addition*, make up that number, as 4, 4, 4, 4, 8. See Plate vii. No. 72. &c. It may therefore be necessary to attend to this distinction in reading the numbers and characters. They had one peculiarity in common with the Romans, which was, that a distinct unit, placed before any number whatever, except an unit, diminished its value. See Plate vii. No. 92, &c.

For the numerical value of most of the characters, simple and compound, consult Plate ix. which may also assist in developing their respective places and powers as alphabetical characters, particularly the simple ones; for the compound characters formed no part of the alphabet, as far as I can judge from such hieroglyphic inscriptions as I have hitherto met with.

Thus, the figure of a bird in numerals is 20; but as an alphabetic character, it seems to represent the letter S or Sh, as appears on the sarcophagus of Alexander the Great, in the name Alexander, which is written *Alegsander;* in the word *Shere,* son, in the same; and in the word *Soth,* one of the names of Mercury, or

Thoth.—Thoth has a symbolic mark, and the name Thoth expressed under it by two semicircles with the arch uppermost, which makes the character *Th;* and representing 8 in numbers, it should stand in the 8th place in the alphabet. The Serpent, which represents 4, is the last in the word *Phre,* a name of the sun, and written under it: therefore *e* is the 4th character in the alphabet. Arranging the alphabet as nearly as possible with the Hebrew, Arabic, Coptic, and other alphabets of the East, throwing together such characters as agree nearly in sound, so as to take in the three characters above mentioned in their proper places, as marked by their numerical values, we shall have the alphabet run in the following form, or nearly so:

1	2	3	4	5	6	7	8	9	10	11	12	13
A	G	D	E	B. F.V.	Z. SZ.	H. E.	TH.	J	K	L	M	N
14	15	16	17	18	19	20	21	22	23	24	25	26
X	O. A. HH	P	TS	Q	R	S.SH.	T	U	PH	CH	PT	O

I do not by any means intend to say that this arrangement is correct throughout: for in the latter part we have no simple characters that I know of, that correspond to the numbers.

In the Sarcophagus of Alexander the Great, so called, we can obtain the powers of several characters. In the name *Philip* we get the two first characters sitting opposite each other, the one *Ph,* the other *L;* and the latter we find also the second in the name, *Alexander.* The character for *p* in Philip is variously written, and the remainder of the name seems to be a title, as it is found with other names. In the name *Alexander,* all the characters are sufficiently plain: after the characters for *A, l,* comes that of a hand and arm for *g;* then the bird, as *s* or *sh;* then a figure somewhat oval, divided by a bar lengthwise, for *n;* then the semicircle with the arc upwards, which is *th;* and lastly, a circle with a cross in it for *r,* making the name *Algshnthr.* The characters are probably syllabic, that is, each carries its own short vowel. The circle for *r* is also found in the word *Sheri,* already alluded to, which word is represented by the figure of a bird, *Sh,* and a circle, *r,* placed between the names of Philip and Alexander; as much as to say, Philip's son Alexander. I mention this merely as a hint to others; for I have not materials sufficient to enable me to enter more fully into the subject. But these helps go but a short way in assisting us; for it seems that the very same letter of the alphabet may be represented ten or a dozen different ways: we see also that the same number is represented by several distinct characters. See Plate ix. This circumstance in itself, exclusive of contractions, mere symbols, &c. must create considerable difficulties. The triple would, I suppose, be sufficient to give a pretty correct view of the hieroglyphic characters; but unfortunately I have never seen but the Greek part, which alone is of no use in the investigation.

NINTH. *Figures that represent Beginnings and Endings.*

The beginning of any thing is represented by a head, and sometimes by a man standing erect.

The endings are marked by figures opposite to the former, viz. by a pair of legs and feet, and by a man sitting.

Thus, for an example of both beginning and ending:—In the calendar, Pl. vii. nearly under No. 22, is the figure of a head and a pair of legs. The day is that of full moon, which in the year 708 of Rome fell on the 13th of February, being the day held sacred to Jupiter Ammon, who is figured in the boat underneath with four rams' heads. Now, if we look into the Roman calendar, we shall find that the 13th of February was called the Ides of that month, which were reckoned backward to the Nones, which fell on the 5th. Therefore from the Ides to the Nones are eight days, which are placed under the figure of the head, to imply that from the beginning of the Ides to their end, marked by the legs, are eight days. The date marked in the calendar of Dendera under the figure of the full moon, and over the heads of Jupiter Ammon, is 13, that is, 4 and 9. (See Plate ix.) Hence it is proved to be only a hieroglyphic translation of the Roman calendar, which gives the same date.

Calendar of the Portico, Plate VII.

We shall begin with this calendar, as the most simple, distinct, and easiest to be understood of the two; which being explained, the other (the circular one, Pl. viii.) will thereby become less difficult, as there are a great many figures in both that refer to the same subject.

The Calendar of the Portico, as it is called, is divided into three distinct columns, the first and undermost representing a number of boats, arranged in regular order from beginning to end. This range of boats seems to have been partly intended for ornament, and partly for the purpose of receiving such figures as could not be represented in the third or upper column for want of room. The second or middle column next to the boats is narrow, and therefore affords room but for few figures: it extends the whole length, and is principally intended to have the date set down in it where necessary, together with marks of the sun's ingress into the signs, and the marks which represent the beginning and ending of each month, &c. The third or upper column contains a variety of figures similar to those in the boats, each having a tablet, on which the day of the month was originally inscribed; but now many of them are defaced by time. Besides these, there are also the figures of zodiacal constellations, which, I believe, from the want of proportion and other circumstances, are only meant to represent the signs.

The calendar, being to long to be represented in one piece, is is divided into two separate parts, each of which is placed on the opposite plat bands of the Portico, with a figure of Isis, or the year, under each, as represented, Pl. vii. The year 708 (of Rome) which is marked on the calendar, was that of the Julian correction, or, as some have called it, the year of confusion; because the ancients, not comprehending the nature of the correction, imagined that Julius Cæsar had added 79 days to the year, making it to contain 444 days, than which there could not be a greater mistake: for Julius Cæsar did not add an hour to the year, which will be seen, if his mode of correction be properly understood. The matter was simply this: He found the whole of the ancient festivals completely deranged; and hence, to correct these, as well as other matters relating to the calendar, he directed that the 15th of October in Numa's year should be called the first of January, at the same time adding one day to the days in November, which before were only 29, and two days to those in December, which in Numa's year were only 29, thereby making the former month to contain 30, and the latter 31 days. Now, it must be obvious, that by the method he thus adopted, he only changed the *name* of the time, without adding an hour, and that the positions of the sun, moon, and planets on the 15th of October in Numa's year, were the same with their positions on the 1st January in the Julian year, allowing for three days that were added. So that the difference of the time was merely nominal, though taken as real by the ancient writers on the subject; for they took it in a different light, and, computing from the 15th of October to the end of that month, they found it 17 days, to which adding November 30 days, and December 31 days, made to the end of December 78 days;

and one day more, to make it the first of January, made the num-79 days from the 15th of October to the 1st of January, both days inclusive, which they supposed to be in addition, and therefore made the year 444 days, that is, 365+79=444. But if Julius Cæsar had added 79 days, as was thus supposed, it would have made no alteration whatever in Numa's year, because then the 1st of January, instead of being on the 15th of October, would fall at its usual time, and no change take place except the three days added in November and December. Now in the calendars of Dendera, we have the same thing expressed or represented on the shoulders and body of Isis, thus:—On the shoulder of Isis, near the top, (see the plate,) are marked on one side 7 units, and on the opposite, 8 units more, making together 15, the day of the month, as above mentioned, at which the correction was made;—next below these are four stars on each shoulder, making together 8, the number expressing the name of the month, October, the 8th month of the year, reckoning from March as the first. Therefore the 15th of October is the day given by the figure of Isis. Then the next thing is to find the addition. Along the right-hand figure of Isis, which begins the calendar, there are a number of stars marked: these being counted down to the ankle-band, give 74, to which adding the five stars in the ankle-band, we have the whole number, 79 days, as the nominal difference to be added to the 15th of October, to find the time in the Julian year, which agrees with what I have above stated. On the opposite, or left-hand Isis, there are marked in the same manner down to the ankle-band 83 stars, which with the three on the ankle-band make 86, the number of days from the first of October, to the winter solstice, both days inclusive, thus: October complete 31 days, November 30 days, and 25th December, make all together 86 days, as marked on Isis. The object in putting three of the stars on the ankle-band, was to point out the three days that were added at the end of the year to November and December; and the five on the opposite band was to remind the Egyptians, that they were to add five days to the end of every year, in order to make it agree or keep pace with the Roman year: and in those years that were leap-years, they were to add one day at the end of the month Mesori, by which means the month Thoth would always begin on the 29th of August, then sextilis, as it did in the year of the correction, and marked so in the calendar, to the translation and explanation of which we shall now proceed.

Translation of the Calendar of the Portico, Plate VII. with explanatory Remarks.

No.	Day of the Month.	
		DATE, THE YEAR 708 OF ROME. This is marked at the head of the middle column, and read from left to right, the last figure of which is directly under No. 1.
1		The Year contains 366 days. This is represented by a small figure of Isis, the emblem of the year. She has a head-dress on of the African or Nubian speckled hen, to denote the vicissitudes of seasons; and a tablet before her, containing the number 365 and a separate unit, to express that the year 708, over the last figure of which she stands, contains 366 days.
		The 30th day of the moon and conjunction. This is expressed by the figure of a bird, which stands for 20, with the figure of 10 under his feet, making 30; and the bird Ibis, the symbol of the moon, next to which is the mark of conjunction.
2		The sun enters the sign the 16th. This is represented by the figure of a man with a hawk's head, a symbol of the sun, with a tablet before him, expressing the date twice in different characters: first 12, 4 = 16; and 4, 4, 8 = 16.
3		January, the 1st month of the year, the sun enters the sign the 25th. This is represented by the figure of Anubis with a crescent on his head, having a tablet with a bird, 20, on it, a serpent, 4, and an eye, 1, making in all 25, the day of the month.
		The sun enters the sign Aquarius on the 16th and 25th January. This is marked by the figure of ingress into the sign, placed between Nos. 2 and 3, and followed by two rows of figures: the under ones are 8, 4, 4 = 16; the upper ones are 12, 13 = 25.
4, 5, 6	1	The sun and moon in conjunction at night on the first of the month, and commencement of the year. This is represented by the figure No. 4, symbol of the sun, standing on the back of a goose, No. 5, symbol of night, or perhaps of Rome; and No. 6, the figure of a man with the head of an Ibis, symbol of the moon, in the boat: all three in a line, with an unit, a head, and the figure of conjunction, in the middle column, to express the first, and commencement of the month and year at such conjunction. In the circular calendar, the owl represents the time as night: here the bird is a goose, which however, with the Egyptians, might represent the same thing: unless we consider it as the symbol of Rome, or as commencement of the Roman year.

No.	Day of the Month.	
7	8	Sacrifices to Janus, on the 8th. This is represented by the figure of a man holding a kid, or some such thing, for the sacrifice, marked underneath by the figure of 8.
8	9	The Agonalia, on the 9th. This is represented by the figure of a man immolated, marked the 9th by the figures 4, 5.
9		A festival for three days, marked in the tablet 18th, but not in the Roman calendar.
10		The sun enters the sign the 25th. Marked by the figure of a female with a tablet, with the number 25 on it, represented by a serpentine figure, 24, and an unit under=25: the female also points to the numbers in the middle column, being 5, 8, and 12.
11		The sun enters the sign the 16th. Marked by a female, but the tablet date not legible.
12		The figure of Aquarius.
		The end of the month of January. Represented by a pair of legs in the middle column, under the figure of Aquarius.
		FEBRUARY.
13	9	The beginning of spring. See the table of seasons under B. Marked by the figure of a man in a boat, with a tablet, with the number 10—1=9, and a head, to express commencement: in the middle column it is also marked 10—1=9.
		A bird of passage marked in the middle column, but name unknown.
14		The sun enters the sign Pisces the 24th.
		The number of days in February 29. Marked by a bird=20, and 9 under it=29, middle column,
15		The sun enters the sign the 17th day. Marked by a female and tablet, the numbers 12, 5=17.
		The number of days in the month of February 29, again repeated.
16		The middle of spring the 24th. See table of seasons, A. Represented by a figure of Anubis, with a tablet, marked 5, 7, 8, 4=24.
		February ends the 29th. Represented by the figure of a small animal couchant, followed by two compound numbers, each equal to 12, and a figure of 5=29.
17		The figure of Pisces.
18		A festival of four days on the 8th, but not marked in the Roman calendar. Represented by the figure, symbol of the sun, in a boat, with a tablet, and the figure of 8 over a star, that is to say, the 8th day.

No.	Day of the Month.	
		The days in February again marked in the middle column=29.
19	11	The Genialick games. Represented by a Phallus under No. 19 in the middle column.
20	11	Arctophylax rises. Represented by the figure of a bear, or bearward, in the middle column, the date 8 and 3=11.
21	13	To Jupiter Ammon and Faunus, the 13th. Represented by the figure of a man in a boat, with four rams' heads, and with two stars above them, to denote that the festival lasts two days: over the heads is the date in the middle column, which is 3, 1, 9=13th, and 20 days after the sun's entrance into Aquarius by the tablet.
22		Full moon the 13th. Represented by the figure of man holding a hog by the hind part within the moon's disc, in allusion to the old legend of the man or woman in the moon holding something, supposed by fancy to be represented by the dark spots. The date is marked underneath, being the same with the last, answering for both.
		The Ides of February, the 13th. The Ides fall on the same day as the last, and as they are counted backwards to the None, which is the 5th; they contain 8 days, represented by a head placed on the figure of 8, and then a pair of legs, to express that from beginning to end they contain 8 days.
23	15	The Lupercalia. Represented by the figures of two boys running, being a part of the ceremonies used in that festival by the Romans.
24		The sun in the sign Pisces, the 15th. Represented by a figure of the sun, with a tablet marked 3, 1, 10, 1=15. The Roman calendar says 16th, but Columella says the 15th.
	23	The swallow appears the 23rd. Represented by a bird in the middle column.
25		The beginning of spring? See table of seasons, C. Represented by the figure of Anubis, with a tablet, date not perfect.
		MARCH.
26		The sun begins to enter the sign Aries the 18th. Represented by a female figure with a tablet, in which is marked in numbers 6, 6, 4+1, 1=18, A bird of passage appears in the middle column.
27		The sun enters Aries on the 25th. Represented by a female figure without a tablet, standing above the date 5, 5, 5, 10=25 and 5, 5, 5, 5, 5=25, in the middle column.
28	12	The beginning of spring. See table of seasons under D. Represented by the figure of Harpocrates sitting on a flower in a boat, having a tablet marked 12th day, with a head, to express commencement.

No.	Day of the Month.	
29		The figure of Aries and middle of spring. Represented by the figure of Aries, or the Ram, surrounded by 45 stars, to denote, that when the sun enters that sign, it is the middle of spring, that is, 45 days from the commencement of that season on the 9th of February; therefore the middle of spring is the 25th March. See table of seasons under B.
30		The middle of spring, the 25th March. Represented by the figure of Anubis.
31	14	Full moon the 14th. Represented on the head of Harpocrates. The tablet marks 4, 10=14, or 6 and 8=14.
32		The festival, second Equiria, upon the Tiber? Represented by the figure of a man with a staff: the date in the middle column is 8, 6=14.
33	17	The Agonalia, and Milvius sets, the 17th. Represented by the animals for sacrifice tied together, surmounted by the bird Milvius: the date underneath is 8, 9=17.
		The end of March, represented by a boundary pillar, or barrier.
		APRIL.
34		The sun enters the sign Taurus the 24th. Represented by a female and tablet, marked 8, 4, 4, 8–24.
35		The sun enters the sign on the 19th, by the Roman calendar. Represented by a female figure—the date in the tablet not legible. Columella says the sun enters the sign the 17th.
36		The sun enters the sign the 24th. Represented by the figure of a man in a boat, with a tablet expressing 8, 13, 2, 2–1=24.
37		The sun enters the sign, and middle of spring the 24th. See table of seasons under D. Represented by the united figure of the sun and Anubis: the tablet illegible, but underneath, in the middle column, is the usual figure, marking the ingress of the sun into the sign, where also the date is 4, 8, 8, 4=24. It seems to have been a festival of four days, by the stars on the Bull's tail.
38	11	The beginning of spring on the 11th April. See table of seasons under A. The numbers in the middle column, being 7 and 4=11, with the figure of a man standing, which signifies commencement.
39	11	Full moon on the 11th. Represented on the shoulders of the Bull.
40	28	The commencement of the Floralia. Represented by the numbers in the middle column over the head of the figure in the boat: the numbers are 9, 8, 10, 1=28, and six stars, 3 behind and 3 before. The figure seems to point by his tablet the number of

No.	Day of the Month.	
		days in the month = 30, or else a festival on that day, to continue four days.
41		The Floralia, for six days. Represented by the figure of a man holding a serpent, with three stars before the serpent, and three after it, and with a flower under his right foot. The serpent denotes the commencement of the year with the month of May.
	30	The end of April, marked by the figure of a man sitting, as a pause or rest.
		MAY.
42	1	The commencement of the year (mercantile) on this day. Represented by the figure of a serpent, held in the hand of No. 41.
	3	The last day of the Floralia. Marked by the figure 3, under the last three stars.
43		The sun enters the sign. Represented, as usual, by a female figure, but her table not legible.
44	15	The birth of Mercury. Represented by a figure with a cap on: the tablet states the date as 8, 8—1 = 15, and the figure has the same number of stars about it.
45	19	The sun enters the sign on the 19th. Represented by a female figure, as usual, her tablet marking 9, 9, 1 = 19th day.
46		An attendant on figure 48.
47	24	Midsummer day, the 24th. See table of seasons, under A. Represented by a female Anubis in a boat, with a small serpent on her head, to denote that it is a particular stated time of the year for settling accounts, making contracts, &c. &c. Her tablet not legible, but over her head in the middle column are the numbers, 4, 5, 5, 5, 5 = 24.
48		The sign Gemini, or the Twins. Represented by two figures; one a male Anubis, with a feather in his cap; the other a female Anubis, holding each other by the hand, and having a number of stars between and about them, equal to 24, to denote that the sun is in Gemini on that day. The female Anubis is the same with the one in the boat, No. 47, representing Midsummer day.
	26	New moon. Represented on the head of the female Anubis, No. 43.
49		An attendant on the female Anubis No. 48, the same as No. 46 is an attendant on the male figure 48: they appear to have no other meaning.
50		It is uncertain to what this figure refers, neither the number above its head in the middle column, nor the remains of his tablet being sufficiently legible.
		The end of May, marked by a man sitting, as usual.

No.	Day of the Month.	
		JUNE.
51		The sun enters the sign Cancer the 19th and 25th. Represented by a figure in a boat, the numbers in the tablet not very legible. Underneath in the middle column are the marks representing the sun and moon's ingress into the sign at the same time, consequently a conjunction, on the 25th of June.
52	25	The sun in Cancer. Marked by the figure of a hawk or eagle, elevated on a perch, to denote that it has attained the highest point in its course.
53		New moon. Represented by the figure of a man with the head of an Ibis in a boat: the date in his tablet imperfect, but the middle column marks 25. The eagle, No, 52, and this are in a line.
		The end of June, marked by the figure of a leg.
		QUINTILIS, OR JULY.
54	8	The middle of Capricorn sets. Represented by the figure of a couchant goat, very ill drawn, in a boat. The six stars or days marked, probably allude to other festivals, the days for which are marked under the boat; as the Poplifugium on the 5th, the Ludi Apollinares on the 6th, the Nonæ Caprotinæ on the 7th, &c.
55	25	The new moon and sun enter Leo. Represented by the figure of a man with a hawk's head surmounted by a figure of the moon: the tablet not very plain, but the middle column marks 10, 6, 5, 4=25, and again 10, 7, 8=25.
56		The middle of Aquarius sets. Represented by the figure of Aquarius in a boat: the date marked is in two rows of figures thus, 5, 5, 1, 1, 1, 1, 3, 8=25. There is another figure in the same boat, to represent the entire setting of Aquarius on the 13th of the following month: the date is 8, 1, 4=13, or 13th of August. See the circular calendar, where the figure of Aquarius is given under Leo, along with that of Diana, the 13th of August.
57		Diana, but here misplaced, being marked in the middle column the 13th of August. Represented by three figures of that goddess, two of them seated, with the moon on their heads; but in consequence of their being here misplaced, the numbers on their tablets were entirely erased: their proper place being under the hind part of Leo, at or behind No. 62, which is the 13th of August, the day sacred to Diana. Aquarius, most probably, had been placed under the hind part of Leo, but erased from thence, to place it where it now stands; and in the removal of Aquarius, that of Diana took place, by mistake. The date in the middle column between Aquarius and Diana, is the 13th of August, thus: first a bird, implying month, or time; then an oval, 5, which with three units annexed, make 8, the number of the month. Within the oval is the

No.	Day of the Month.	
		date 12 and 1 united, to express that it is the 13th day of the 8th month, reckoning from January as the first.
58		This is the same with No. 62, here misplaced, along with Diana.
59		The sun enters Leo the 20th, by the Roman calendar. Represented by a female figure as usual, standing at or near the head of the serpent Hydra: the numbers in her tablet not legible.
60		The sun and moon enter Leo. Represented by a figure of the sun, and moon on its head. This is the same with No. 55 above given, and here again repeated, which shows that there is considerable confusion in this part of the calendar. The date is 25th July as before, viz. 9, 1, 5, 10=25.
		The end of July marked by a pair of legs.
		SEXTILIS, OR AUGUST.
61	4	The middle of Leo rises. Represented by the figure of Leo, having his left fore paw nearly over the date 4.
62	13	The feast of servant maids and slaves, and the Ides of August. Represented by a female in a boat, with a serpent: the latter seems to express that this is some particular period in the year, with the nature of which I am not acquainted. The Ides are expressed by the figure of a head in the middle column: the date marked in the tablet is 8, 1, 4=13th of August. This is the place where the figures of Diana, No. 57, should come in.
63	24	The beginning of Autumn. See table of seasons, under C. Represented by the figure of a man in a boat, the tablet not legible; and the figure of a small head in the middle column, to denote commencement. A three days' festival.
64		The sun enters Virgo the 24th. Represented by a figure in a boat, having a pot on his head: the tablet date is 8, 8, 3, 5=24. A three days' festival.
65		The sun enters the sign Virgo the 20th. Represented by a female, the numbers on whose tablet are imperfect: but under her feet in the middle column is marked 8, 12=20.
66		The sun enters the sign the 24th. Represented by a female figure, her tablet expressing the following numbers, viz. 4, 7, 5, 8=24: the same expressed differently by the numbers under his feet, as 6, 8, 6, 4=24.
67		The figure of Virgo requires no explanation.
68		The sun enters Virgo the 20th. Represented by a figure in the boat, nearly under Virgo. The tablet numbers incomplete, but in the Circular Calendar, the same is marked 10, 1, 9=20.

No.	Day of the Month.	
69		**The commencement of the month Thoth, the first of the Egyptian year.** Represented by a figure of the Anubis kind, with a crescent and a star on its head, and a torch at his foremost foot, to denote that it commences the year. Under the same foot are marked in the middle column three tens, to denote that the month contains 30 days: near which, and directly under a bird with a dog's head, is the figure of a small head, to signify commencement; and under the head is a tablet over the boat, in which the date is inserted, viz. 10, 10, 9 = 29, or the 29th of August.
70		**The commencement of Thoth** Is again represented by the figure of a bird with a dog's head, or time, united to Anubis. It is directly over the small head in the middle column, and the tablet over the boat, as already mention d.
71	30	**First quarter of the moon the 30th.** Represented on the head of the figure in the boat, which has a star between it and the tablet, to express that the day is one day later than that date, making the first quarter on the 30th. It has also two stars after it, to denote that two days after the first quarter is the first of September, marked in the middle column by the figure of a small head.
		SEPTEMBER.
	1	**The beginning of the month.** Marked by a small head in the middle column, as mentioned above.
72		**The sun enters the sign Libra the 24th.** Represented by a female as usual. The tablet date is 8, 4, 4, 4, 4 = 24.
73		**The sun enters the sign Libra the 19th.** Represented by a female as usual. The tablet date is 8, 4, 3, 3, 1 = 19.
74		**The sun in Libra the 24th.** Represented by the figure of the balance and Harpocrates under it, within the figure, which usually marks ingress of the sun. The date is repeated in the middle column in different numbers and characters.
75	21	**New moon the 21st.** Represented by the moon being placed on the head of a man in a boat below the balance. The tablet date is 20, 1 = 21st day.
76		**This figure marks the entrance of the sun into the sign the 24th.** The tablet is 20, 4 = the 24th day.
77	29	**First quarter of the moon the 29th.** Represented by a circle with a female within it, and the bird symbol of the moon standing on top of it. The numbers in the middle column are here erased, as if some mistake had arisen; but from the numbers in the tablet below = 24, this must be the 29th of the month.
		This ends September. The usual mark for termination is erased.

No.	Day of the Month.	
		OCTOBER.
78	11	The beginning of winter. Represented by the figure of Anubis; but the dates are all erased from the middle column, and therefore we must be guided by those in the boats below.
79		The sun enters Scorpio. Represented by the figure of a female as usual; but there is no tablet nor date to refer to.
80	13	Last quarter of the moon the 13th. Represented by a figure in the boat, having the moon on its head. The tablet date is 8, 1, 3, 1 = 13.
81		The sun enters the sign the 20th. Represented by a figure of the Anubis kind, under whose tail is the date 1, 12, 5, 1, 1 = 20.
82	21	New moon the 21st. Represented by the moon being placed on the top of a support, held by a figure in the boat. This is a figure of the sun in autumn, pointing out the season for threshing the corn, of which more may be seen in the Circular Calendar.
83		The figure of Scorpio, which requires no further explanation.
84	28	The commencement of the Egyptian month Athyr. Represented by the figure of a bird, to denote the sun's entrance into the month Athyr; under which, in the middle column is the figure of a small head.
85		The same otherwise. Represented by the figure of a wolf, with a small serpent and a torch, to express that it was the commencement of the year at some former period. The date marked in the middle column is 20, 4, 4 = 28.
86		Represents a festival for five days, by the number of stars. The hand implies two, and the head one, making up the number three, to denote that Athyr was the third month of the Egyptian year. The three posts erected in the boat also denote the same.
		NOVEMBER.
87	5	Full moon and eclipse. Represented by the figure of the moon on a boat, with Anubis on the disc. The date marked in the middle column is 5, and in the tablet is 3, 1, 1 = 5, with a head to express commencement, and an eye and a bird above, as additional symbols of the moon. Some birds of passage appear also in the middle column, ready to depart, and one in the act of flying; but we know not their names.
88		The sun enters the sign Saggitarius the 24th. Represented by a female figure and tablet, marked 4, 12, 4, 4 = 24.
89		The sun enters the sign the 18th. Represented by a female figure and tablet, marked 8, 5, 5 = 18.

No.	Day of the Month.	
90		**The sun enters the sign the 24th.** Represented by a figure of the sun in a boat, with a tablet marked in two rows 6, 5, 1 and 4, 6, 1, 1 =24.
91		**The sun enters the sign the 18th.** Represented by the figure of Saggitarius, with the mark of ingress into the sign underneath, and the figures 8, 10 =18. Thus we may perceive how often the same thing is repeated, as if with a view to prevent mistake, or to make it more clear or better understood to different persons, who might understand one mode, and not another. The bird on the hind part of Sagittary represents the new moon on the 19th; and the Anubis united with Sagittary, marks the beginning of winter. See seasons, under C.
92		**New moon the 19th.** Represented by the figure of the moon, placed on the head of the symbol of the sun in a boat. The tablet marks—1, 20 =19th. Here we have an instance of the number, when preceded by an unit, being diminished in value, as in the Roman method.
93		**The sun enters the sign the 24th.** Represented by a female figure, pointing to the date in the middle column 20 and 4 =24, which differs from the others already mentioned, this being a bird =20, and a broken chain with four links =24 the whole.
94 } 95 }		**The horns of the Bull set the 20th.** Represented by the figure of a bull on one leg, No. 95, opposed by a figure of the sun, No. 94, holding a dart against one of the horns, as it were to stay their setting.
96		**Uncertain — but a festival of two days.** Represented by a figure in a boat, having a small serpent on its head, denoting therefore some particular day. The tablet not complete, without which it cannot be read.
		The end of November marked by a man sitting.
		DECEMBER.
97	12	**The Agonalia for three days, the 11th, 12th and 13th. and the moon in her last quarter the 12th.** Represented by the figure of a man in a boat, with the head off, and its place supplied by the figure of the moon. The festival lasts three days, as marked by the three stars. The date, as expressed in the tablet, is 5, 4, 3 =12th day.
98		**The beginning of winter. See table of seasons, under D.** Represented by a female Anubis, generally called Nephthe, and mother of Anubis; but this distinction is not of the slightest consequence, for any of the family answers the same purpose to represent time. She holds the bull, No. 95, by a chain, the number of links of which express the day of the month, with a star at the end of it. Properly drawn, it should contain 12 links, which with the star denotes the 12th day.

No.	Day of the Month.	
99	18	The sun enters Capricorn the 18th. Represented by a female figure with a tablet, the date marked in which is 4, 14 = 18th day. The 14 is a compound number, made up of 6, 1, 5, 1, 1 = 14.
100	25	The winter solstice the 25th. Represented by the figure of Capricorn, with the sun, under the symbol of the eagle on the top of its horns, to denote that it has attained its utmost limit in its course. The date marked underneath is 6, 6, 6, 6, 1 = 25th day.
101		The sun enters the sign the 25th. Represented by a figure, symbol of the sun, in a boat: tablet marked 25th.
102		The same. Represented by the figure, of a female, the tablet numbers not legible; but she points to the date in the middle column, which is 10, 8, 3, 4 = 25th day.
103		The end of December, and of the year. Represented by a sitting figure, as the usual mark of end, termination, cessation, &c.

We shall now proceed to the Circular Calendar, which, from the irregular manner in which it has been drawn and sculptured, it is very difficult to arrange in proper order, notwithstanding the assistance we derive from the above; and in all probability, with all our care, errors in this respect will escape us, as we have but few dates to guide us in the labyrinth.

Translation of the Circular Calendar, Plate VIII. with Explanatory Remarks.

		DATE THE YEAR 708 OF ROME. Month January. Represented by the outer circular space, in a line with Nos. 1, 2, and 3, and having an unit with a small erect serpent above the last figure of the date, 8, to denote it is the first point or commencement of the year.
1, 2, 3	1	Conjunction of the sun and moon at midnight. Represented by the figures No. 1, symbol of the moon, No. 2, symbol of the sun, and No. 3, the owl, symbol of night, all in a straight line; the figures for the sun and moon being each marked with a star and an unit above it, to represent the first day of the year, the first of the month, and the first of the new moon. The new moon is also indicated by the representation of a pair of eyes, near the date in the outer circular space.
4	8	Sacrifices to Janus. Represented by the figure of a man, holding the victim by one hand, and a club in the other, the same as No. 7 in the calendar of the portico, Pl. vii.

No.	Day of the Month.	
5	9	The Agonalia. Represented by the figure of a horse with human feet, without a head. This is the same as No. 8. in Pl. vii. differently exhibited.
6		The bird Ibis, symbol of the moon, represented in No. 7.
7	15	Full moon the 15th. Represented by an oval, enclosing the figures of eight persons kneeling with their hands behind their backs, in allusion perhaps to some old legendary tale respecting the cause of the spots on the moon's face or disc. The date above it is—1, 10, 6=15, the unit being negative by position.
8		A fish over the full moon No. 7. Supposed to be also a symbol of the moon: but see under No. 16.
9		A fish in the outer circular space. Supposed a symbol of the moon, and to refer to either No. 7 or No. 12. See under No. 16.
10	17	The sun in Aquarius—figure of Aquarius.
11		The Ibis, symbol of the moon, on the head of figure No. 12.
12	22	Last quarter of the moon the 22nd. Represented by the figure of a sheep, with the moon on its head.
13	24	The Sun in Aquarius. Represented by a figure next to Aquarius—the date not legible.
14	25	The middle of winter. See table of seasons, under D. Represented by a figure of Anubis, the date 12, 3, 10 =25.
15	30	Fidicula, the harp, sets. Represented by a man with a harp in his hands.
16		A fish, symbol of the moon? No. 8, 9, and 16, from their situations, appear as symbols of the moon. Ovid, in his Fasti, says, the Dolphin rose the 9th January, and set the fourth February; but these do not agree with the times and positions given: therefore I am rather disposed to think them symbols of the Moon, like the Ibis, Nos. 6 and 11.
		FEBRUARY.
17	9	The beginning of spring. Represented by an inverted flower under No. 17.
18		The figure of Pisces, which requires no explanation.
19	13	To Jupiter Ammon and Faunus. See Cal. Portico, No. 21. Represented by a figure on a pedestal having four rams' heads, to denote the sun in Aries, and next to it the figure of Faunus seated: over the heads of Jupiter Ammon is placed the moon, to denote its being then full. To the left of the moon, and nearly over the head of Faunus, are two birds, with a star to the left of them: these

No.	Day of the Month.	
		signify forty days that the festival had deviated or fallen back at the time of the Julian correction from Aries, the position at the time of instituting the festival: therefore forty days being reckoned from Aries, or the 25th March, gave the 13th of February, to which it was fixed for the future. The cause of its falling back, was owing to the Egyptian year being one day deficient in every four years: hence, 40 × 4 = 160, the number of years then elapsed since the festival coincided with Aries; and this was the antiquity of that festival at the time of the Julian correction. A little to the right of the figure of the moon, over the head of Jupiter Ammon, are four birds more, with a star: these signify eighty days, each bird being 20, and relate to the Julian correction already explained; being the 79 days nominally counted from the 15th of October to the 1st of January, both included, and one day more for February, the year being bissextile, make 80, the number here marked.
20		**Full moon the 13th. See Cal. Portico, No. 22.** Represented by the figure of the moon, with that of a female holding a hog, or some such animal by the tail, on the moon's disc, in allusion to some legendary tales respecting the moon's spots. In the other calendar the figure is a male holding a hog by the tail, date 13th.
	16	**The sun in Pisces the 16th.** Represented by a cap, 12, and 4 stuck into it = 16, marked in the circular margin.
	24	**The sun in Pisces the 24th.** Marked in the circular margin by 5, 5, 5, 4, 2, 3 = 24th.
		MARCH.
21		**The sun in Aries the 17th, according to Columella. The Roman calendar says the 18th.** Represented by the figure, symbol of the sun, with the numbers 8, 9 over his head: the number is compound, and made up of 8, 4, 1, 1, 1, 1, 1, in form of a globe.
22	12	**The beginning of spring. See table of seasons under D.** Represented by Harpocrates sitting on a flower, and two figures of Anubis above him. The numbers appear incomplete.
23	14	**Full moon the 14th.** Represented by the figure of the moon, with an eye in the centre.
24	17	**The Agonalia, and Milvius sets the 17th.** Represented by the victims tied together, surmounted by the bird Milvius.
25		**The constellation Aries sets?** Represented by a figure over the heads of the Anubises, resembling a sheep: but as a sheep is a symbol of the moon, it may represent the last quarter of the moon, placed on the head of the ram, No. 27.
26		**The quadriennium of Julius Cæsar.** Represented by four small serpents erect on their tails on a square pedestal, each of which signify 360; and

No.	Day of the Month.	
		over the head of each is a small circle, which stands for 5, with a star over each, to represent 365 days: but over the head of the first serpent to the right, there is an additional star, to denote that the first year of the four, or of the Julian correction, was a leap year, and contained 366 days.
		The sun enters Aries the 24th. Represented by the number in the circular margin, being 6, 8, 10.
	18	**The sun enters the sign the 18th.** Represented by the numbers in the margin, but not complete.
27	24	**The vernal equinox.** Represented by the figure of a ram's head and neck in a boat. The numbers below are 5, 5, 5, 9=24th: the 9 is compound. Over the head, between the horns, is the symbol of the moon, to denote the third quarter, with three stars to the right, to denote that the time is three days later than the entrance of the sun into Aries on the 18th, or else three days earlier than his entrance on the 24th. Both days may be seen in the Roman Calendar.
28	25	**The middle of Spring. See table of seasons under B.** Represented by a figure of Anubis séated on its legs, having four small serpents over its head, to denote the succession of four years, and seven stars before him, to imply that the middle of spring is seven days after the entrance of the sun into Aries, which, according to the Roman Calendar, was the 18th.
		The end of March, marked by a man sitting in the circular margin.
		APRIL.
29	11	**Full moon the 11th.** Represented under the figure of a boar or hog, being one of those figures marked sometimes on the moon's disc, as in No. 22, Calendar of the Portico. The disc is not always necessary to identify the moon, as will be seen hereafter in another instance, in which it dispensed with; but the figure of a boar here may also be intended to represent the Hyades, which the Romans denominated Sus, Sucula. In the Calendar of the Portico, pl. vii. the full moon is placed on the bull's shoulders. See No. 39.
30		**The sign Taurus.** Represented by the figure of the bull. The sun enters it the 19th, according to the Roman calendar: Columella says the 17th.
	16	**The Hyades set; that is, begin to set.** Represented by the date, 6, 10, over the stars.
	18	**The Hyades hide themselves in the evening. Columella.** Represented by the date, 8, 10, a part of which only is legible.
31	25	**Middle of spring, the latest. See table of seasons under D.** Represented by the figures of two men apparently alike,

No.	Day of the Month.	
		with a date marked underneath them, made up in the form of a cross, consisting of 12, 12, 1=25. Under are various birds of passage marked, with which I am not acquainted: the spotted one seems to denote the new moon.
32	28	Time of the commencement of the Floralia for six days. Represented by the same figure as in the Calendar of the Portico, No. 41, which see. Here the bird seems to represent the new moon on the 26th.
		MAY.
33	1	The first of May, and commencement of the old mercantile year. Represented by the figure of a long serpent in folds, placed on a pedestal, and coronated.
34	15	The birth of Mercury. Represented by the same figure as in the Calendar of the Portico, No. 44, but no date: it has a bird after it.
35		The Twins. Represented by the figures of two men exactly alike, behind No. 32. As there is no date, they require no further explanation here.
36	21	The Agonalia of Janus. Represented by the hind leg of the victim, with some other victims tied to it.
		JUNE.
37	25	The summer solstice. Represented by the figure of an eagle perched on a stand, to denote that the sun is at its utmost limit in its course. This is also midsummer by the table of seasons under B.
38		The sun enters the sign the 25th. Represented by the figure of the sun, with a date 7, 8 10=25. In the marginal space, a little further on, is marked a conjunction of the sun and moon at the summer solstice. In this part of the calendar the figures are intermixed with each other, in such a manner, as to make it difficult to distinguish them with sufficient accuracy.
39		The sign Cancer. Represented by the figure of Cancer; but from the crowded manner in which the figures are placed, it seems to be out of its proper situation.
40		The sun enters Cancer, supposed the 20th. Represented by a figure of the sun, or man with a hawk's head, under Cancer, but no date.
41		The cosmical rising of Sirius. This is supposed to be represented by a small Anubis, or dog, over the back of Leo.
		QUINTILIS, OR JULY.
42	6	The Twins entirely set; and the middle of Cancer sets. Represented by two small figures, holding each other by the hand, near the right claw of the Crab.

No.	Day of the Month.	
43	8	**The middle of Capricorn sets. Roman Calendar. Columella.** Represented by the figure of a goat, in a boat. The figure is very ill drawn, but it can be intended for no other than Capricorn.
44		**The sun enters Leo the 25th.** Represented by the figure or symbol of the sun, the date 15, 10 = 25.
45		**The sun enters Leo the 20th.** Represented by the figure of a man, with the date 6, 4, 10 = 20.
46		**Leo and the Hydra begin to rise cosmically.** Represented by the figure of Leo, with Hydra under him.
47		**The heliacal rising of Sirius.** Supposed to be represented by the figure of a small Anubis, or dog, standing on the tail of Leo.
		SEXTILIS, OR AUGUST.
48		**The sun enters Virgo the 24th.** Represented by the figure of the sun, with the date 6, 5, 4, 9 = 24 ; and underneath in the circular margin 5, 5, 5, 9 = 24 : the 9 is compound, and made up of 4, 1, 1 1, 1, 1 = 9. The three stars imply that it is a festival of as many days.
49	13	**Sacred to Diana.** Represented by the figure of Diana, with a bow and arrow in her hands. Under her is another figure of Diana, with a crescent on her head, and the date 5, 8, = 13th day.
50		**Aquarius entirely sets,** Represented by the figure of Aquarius, with a waterpot in each hand.
51		**The Sun enters Virgo the 24th.** Represented by the figure of the sun, date 20, 4 = 24 : a festival of three days. The date is also marked in the circular margin by 8, 8, 8 = 24, and repeated by 3 other 8's, differently made from the first.
52		**The sun enters Virgo the 20th.** Represented by the figure of a man, date 10, 1, 3, 3, 3 = 20th day. He points to three stars, to show that it is a festival of three days.
53		**Sacred to Isis and Orus.** Represented by the figure of Isis, with the infant Orus standing in her lap.
54		**Middle of the Crow sets.** Represented by the figure of a crow above the head of Orus, and standing on the serpent Hydra.
55		**The sun enters Virgo.** Represented by a figure of a man, above the ear of corn in the hand of Virgo. No date is marked, unless in the circular margin, where they are incomplete.
56		**The sign Virgo.** Represented by the figure of Virgo, and the spike of ripe corn in her hand.

No.	Day of the Month.	
57	23	New moon the 23rd. Represented by the figure of the sun, with the moon on his head, and pointing to five stars, to represent so many days' festival. The date near the figure is three units, which added to 20, the date of the figure next before it, makes 23. The same is marked in the circular margin by a bird, 20, and three units under a star =23rd day.
58	24	The beginning of Autumn. See table of seasons under C. Represented by the figure of Anubis, with a new moon on his head, which, as being one day later than the day of new moon, he points to six days, which is the former five, and one day more.
59	29	The commencement of the Egyptian month Thoth, and of the Egyptian year. Represented by the figure of Thoth, with a scythe in his hands. The crescent on his head implies commencement of the year, which may also be seen in the figure No. 3, Calendar of the Portico. The date underneath is 20, 4, 5 =29.
60	30	The first quarter of the moon. Represented by a figure holding a support in his hands, on which rests an animal like a sheep, and a bird on the top of it, both symbols of the moon, and the commencement of harvest.
61		The commencement of harvest. This is represented by a figure above the last, having a cutting instrument in his hand; and the same is also expressed by the scythe in the hands of the figure representing Thoth: so that the corn in the hand of Virgo was, on the 29th of August, supposed to be sufficiently ripe for cutting.
62		The moon in her first quarter. Represented by the figure of Anubis, with the moon on its head.
		SEPTEMBER.
63	8	Departure of the swallows. Represented by the figure of a bird with expanded wings: date 8th.
64	21	The constellation Pisces sets in the morning. Columella. Represented by the usual figure of wavy lines.
65		Leo entirely sets. Represented by the figure of Leo, under the balance.
66		The sign Libra, or the Balance. Represented by the figure of the balance, with Harpocrates and the wolf, symbols of the sun and the autumnal equinox.
67		The sun enters the sign the 24th. Represented by a man with the date 20, 4 =24th day.
68	29	The moon in her first quarter. Represented by the moon on the head of the sun, sitting at the end of the balance.

No.	Day of the Month	
		OCTOBER.
69	11	The cosmical setting of Sirius. Represented by the figure of a small Anubis, or dog, holding the tail of No. 65.
70	14	The moon in her last quarter. Represented by the moon placed on the head of the sun under the last figure.
71	15	The swallows entirely disappear. Represented by the figure of a bird in the circular margin, with wings expanded; the figure of a man sitting before it, signifying end of appearance.
	21	New moon, indicated by a pair of eyes in the outer circular space, nearly under No. 73.
72		The sun in Scorpio the 24th. Represented by a figure of a man, with a flail on one shoulder and a whip on the other: these appendages imply that it is now time to thresh the corn that has been cut: the whip is to drive the bullocks that tread out the grain. It is properly a figure of the sun in autumn.
73	25	The middle of Autumn the 25th. Represented by the compound date in the broad circular space beneath, made up of 1, 5, 8, 8, 3 =25th.
74	28	The first day of the Egyptian month Athyr, and first quarter of the moon. Represented by the head of a wolf on a pedestal, surmounted by a figure of the moon.
75		The same thing. Represented by the figure of a wolf near the centre of the plate, with a cutting or harvest instrument under his feet, to denote that harvest is entirely over.
76		The sign Scorpio. Represented by the figure of Scorpio, and requires no further explanation.
		NOVEMBER.
77	5	Full moon. Represented by the figure of Anubis in a boat. This figure in the Calendar of the Portico is surrounded with the moon's disc, (see No. 87,) which shows that the moon's disc is not always necessary to represent the moon's phases.
78		The sign Sagittarius. The sun enters it the 18th, and new moon the 19th. Represented by the figure of Sagittarius, with the bird Ibis, a symbol of the moon, on his back.
79	24	The middle of winter. See table of seasons under C. Represented by the figure of Anubis.
80	24	Midwinter the 24th. See table of seasons under C. Represented by another figure of Anubis, next the last.
81		The sun enters Sagittarius the 24th. Represented by a figure of the sun, next the last.

No.	Day of the Month.	
82	29	The Twins begin to rise in the evening. Represented by the figures of two men near each other, and exactly alike, at the back of Sagittarius.
		DECEMBER.
83	12	The Agonalia, and last quarter of the moon. Represented by a figure of a man, with the moon on his shoulders, in place of his head.
84	25	The winter solstice. Represented by the figure of Capricorn.
85		The sun enters the sign. Represented by the figure of the sun, under the last.
86	27	The first quarter of the moon. Represented by the figure of a sheep, with the moon on its head.

This terminates the Circular Calendar; but at the corners of the ceiling of the apartment which contains it, there is given either a supplementary calendar, or rather one of a later date; for the year to which it seems to refer, if I am not mistaken, is 747 of Rome, or 39 years later than those I have just explained. It is, however, on so small a scale, and the numbers so imperfect and indistinct, that it would be no easy matter to make it out. This, however, may be no great loss, as in all probability it is similar to those given; for it begins the rising of Lyra on the 5th of January, and the entrance of the sun into Aquarius on the 17th, &c. I have therefore omitted it in the plate.

Having adverted to the circumstance of the sun entering the same sign twice in every month, which to some may appear extraordinary, as inconsistent with real facts, I shall now give the explanation which I promised, because I do not find that any of the Roman writers, whose works have come into my hands, have given the slightest intimation of the cause, though they mention the fact, and expressed it in the Roman calendar in those months in which the solstices and equinoxes fall.

The cause, however, of this seeming inconsistency is simple, and founded on an ancient custom, that existed long before the time of Julius Cæsar. It is this:—In early times, the months and the signs coincided, and were therefore considered the same; and though in process of time they deviated, yet custom still held them the same, and no alteration was made on that account: so that if the solstice happened to fall on the 8th, 10th, 12th, or 15th day of the month, according to the quantity of deviation, it would be stated as being in the 8th, 10th, 12th, or 15th degree of the sign, by reason that the month and sign were considered as synonymous, or the same: and this practice prevailed down to the time of the Julian correction, which was the cause of the sun appearing to enter the sign twice in the month; once according to the ancient method, the other according to the new, or rather Chaldean, as it appears to have been called. In order to show this more clearly, let us carry back the calculation to the old year of Numa, and see how they will then stand. I have stated above, that the nominal difference between the old and new style was 79 days, which made the 15th of October to be called the 1st January. Now by the Roman calendar, Julian style, the sun is said to enter the sign Aquarius on the 16th of January. On what day, therefore, of the old year of Numa did this fall? If the 15th of October be the 1st of January, then the 16th of January must be the 30th of October, the day on which the sun was supposed to enter the sign in the old year. Again, as the sun entered the sign on the 18th of December by the Roman calendar, on what day of the month did that fall in the old year? From the 18th of December to the 1st of January is 14 days: count these back from the 15th of October, exclusive of that day, because already reckoned, and it will carry you to the 1st of October, the day on which the sun was supposed to enter the sign by the old method. Thus, I think, I have given sufficient proof that the months and the signs were considered the same in ancient times. But to proceed farther:—the winter solstice fell on the 25th of December, according to the calendar of Julius Cæsar, which is the 8th day after the 18th of December, both inclusive; but the 18th of December I have shown to have been the 1st of October, consequently the 25th must be the 8th of October, the day on which the winter solstice fell: and as the month and the sign were considered the same, the solstice of course would be said to be in the 8th degree of the sign,—a circumstance which, though very simple in itself, has perplexed many of our modern astronomers, who could not comprehend how the solstice, which of itself was the beginning of the sign, should be in the 8th degree.

Columella is the only writer that I know of that speaks on this subject, but without entering into any explanation. He says, Book ix. ch. 14. " From the setting of the Pleiades to the winter " solstice, which happens almost about the 23d of December, in " the *eighth degree* of Capricorn, &c. Nor am I ignorant of

"Hipparchus's computation, which teaches, that the solstices and "equinoxes do not happen in the eighth, but in the first degree of "the signs. But in this rural discipline, I now follow the calen- "dars of Eudoxes and Meton, and those of ancient astronomers, "which are adapted to the public sacrifices; because husband- "men are both better acquainted with that old opinion which has "been commonly entertained; nor yet is the niceness and exact- "ness of Hipparchus necessary to the grosser apprehensions and "scanty learning of husbandmen." Again, Book xi. ch. 2. he says: "The 17th of December the sun passes into Capricorn: it "is the winter solstice, as Hipparchus will have it. The 24th of "December is the winter solstice (as the Chaldeans observe)." Here Columella would make a distinction, as to the supposed difference between Hipparchus and the Chaldeans. But he was mistaken in his idea: he supposed the sun entered the sign on the 17th of December, because, according to his opinion, it was the beginning of the month in the old year; but it did not follow from thence that Hipparchus would have made it the solstice: on the contrary, he would, as before expressed, have made the sign to have commenced at the solstice, whether the time of it was by his own observations, or by the Chaldeans, if he found it right.

The explanation now given, while it serves to show the original cause of the two entries of the sun into the same sign in one month, will, I hope, likewise be sufficient to point out the fallacy of depending on the positions of the colures, as mentioned by ancient writers, for determining the antiquity of the time to which they might be supposed to refer; for we here see, that the degrees referred to for the position of the colure are nothing more nor less than the days of the month, which cannot be of any real use, as we are totally unacquainted with the various changes they may have undergone, and the times when such changes were made. Therefore nothing short of an actual observation, referring the positions of the colures to some fixed star, can be relied on.

With respect to my translations of the two calendars above given, I hope they will be received with indulgence, and that weight and attention their apparent correspondence with the Roman calendar entitles them to. All I can now say is, that I have spared no pains to render them as correct as the imperfect state of the originals would allow, and that I have spent a great deal of time in their investigation. Should, however, my labours in this respect ultimately prove useful, in throwing a ray of light on the subject of hieroglyphics, which was the principal object I had in view in undertaking so difficult a task, I shall not deem the time employed as lost or thrown away.

FINIS.

Taylor, Green, & Littlewood, Printers, 15, Old Bailey.

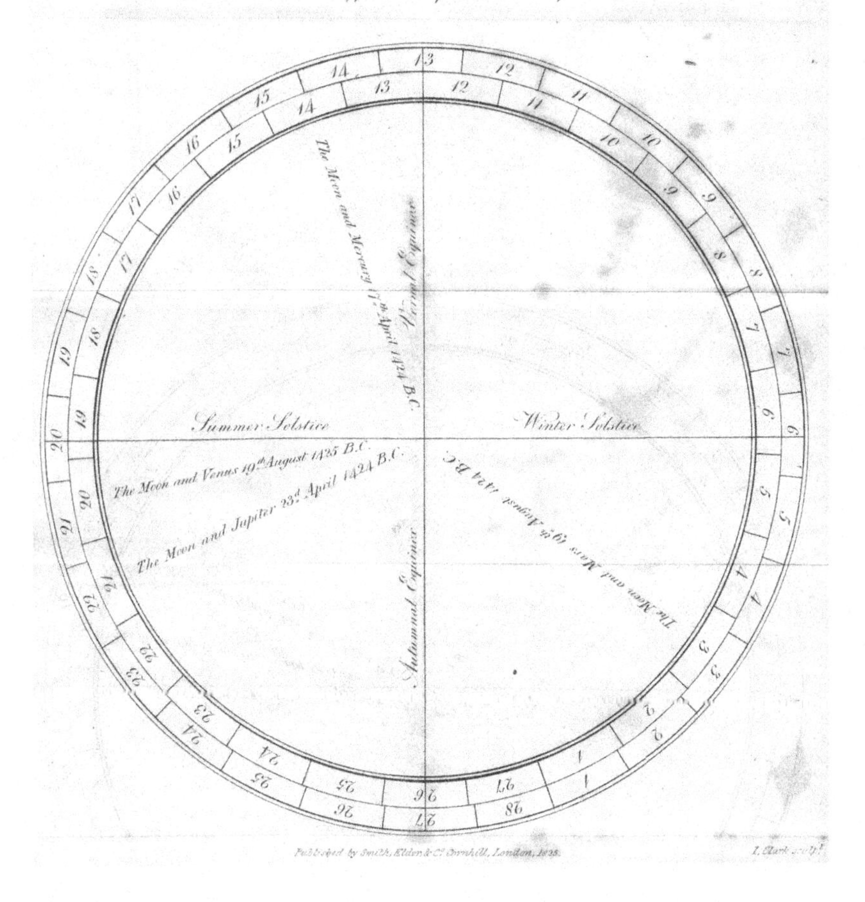

The material originally positioned here is too large for reproduction in this reissue. A PDF can be downloaded from the web address given on page iv of this book, by clicking on 'Resources Available'.

Plate II.

The Relative Positions of the Lunar Mansions, Colures, Months, and Seasons, in 1192 B.C. when the Hindu Months were first formed, and names assigned to them on Astronomical principles, &c.

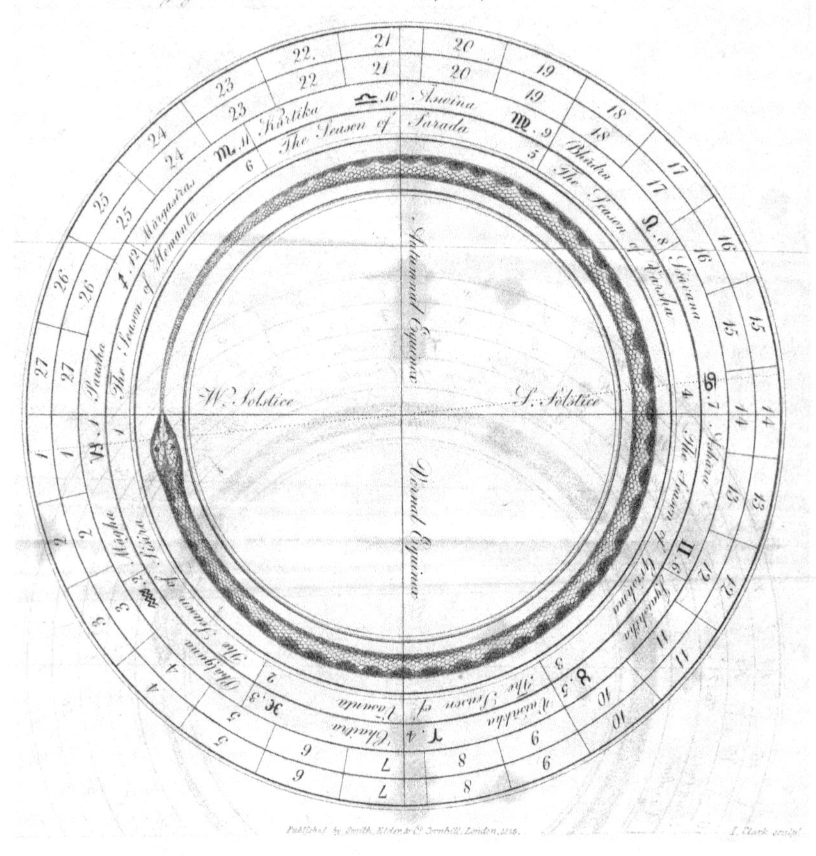

The material originally positioned here is too large for reproduction in this reissue. A PDF can be downloaded from the web address given on page iv of this book, by clicking on 'Resources Available'.

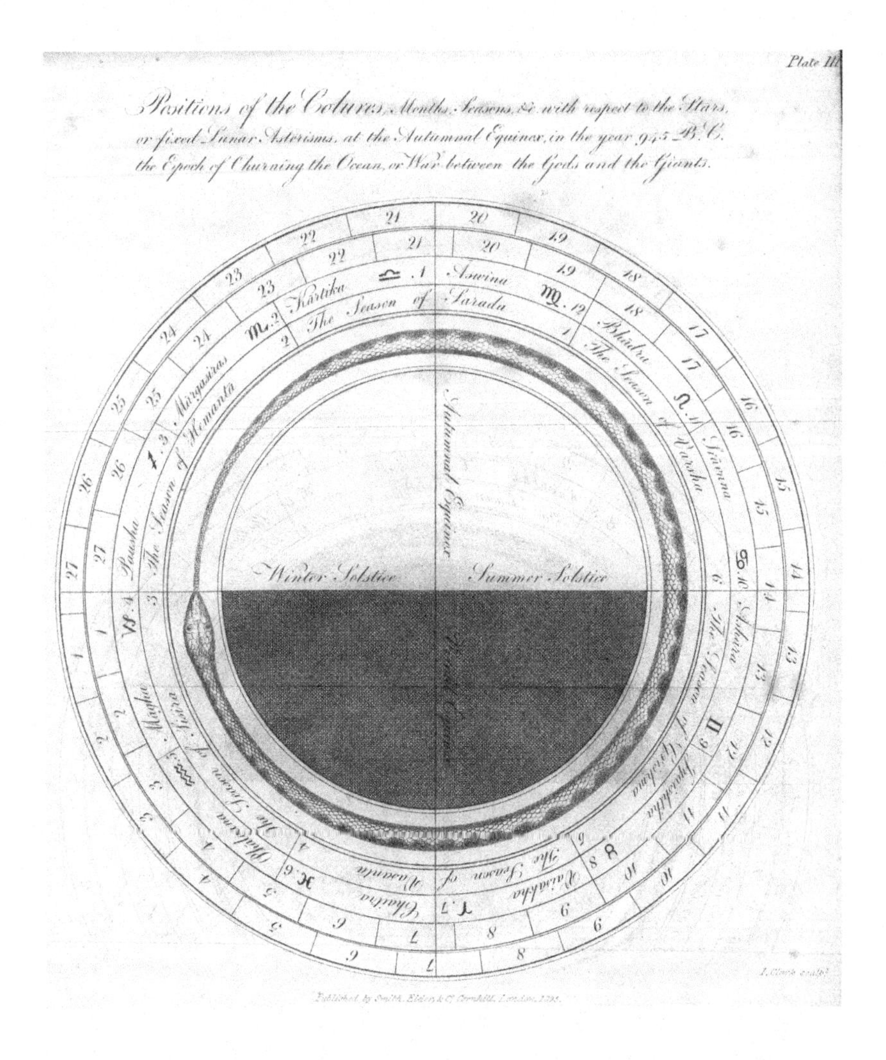

The material originally positioned here is too large for reproduction in this reissue. A PDF can be downloaded from the web address given on page iv of this book, by clicking on 'Resources Available'.

Plate IV.

The Positions of the Colures, one of which passes through the Star δ in Cancer; the coincidence of the Egyptian Months with the Signs, and Cancer and Capricorn partly immersed in the sea, at the War between the Gods and the Giants, in Egypt and the West.

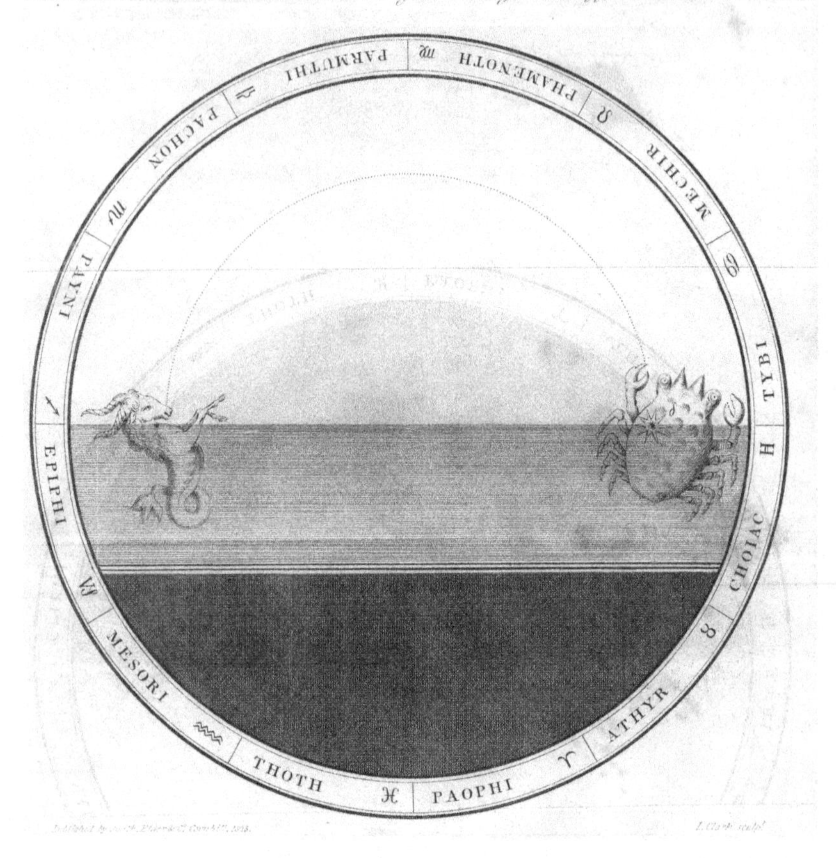

The material originally positioned here is too large for reproduction in this reissue. A PDF can be downloaded from the web address given on page iv of this book, by clicking on 'Resources Available'.

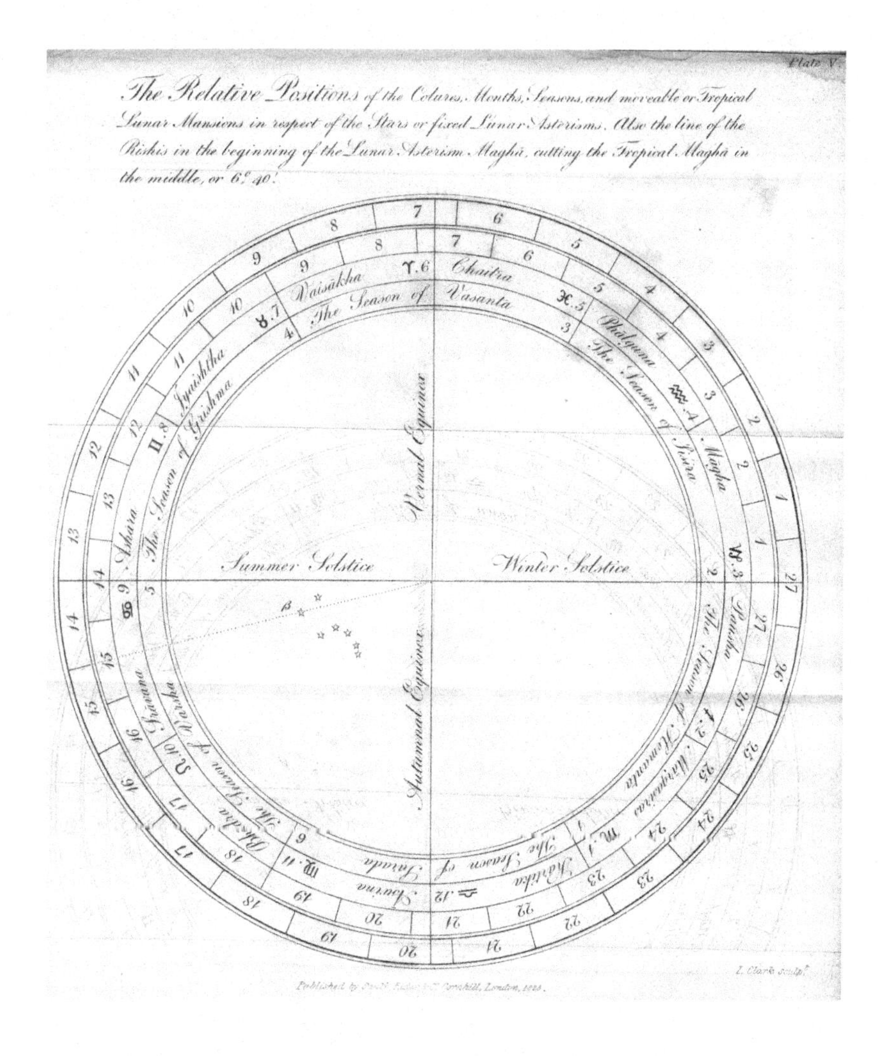

The material originally positioned here is too large for reproduction in this reissue. A PDF can be downloaded from the web address given on page iv of this book, by clicking on 'Resources Available'.

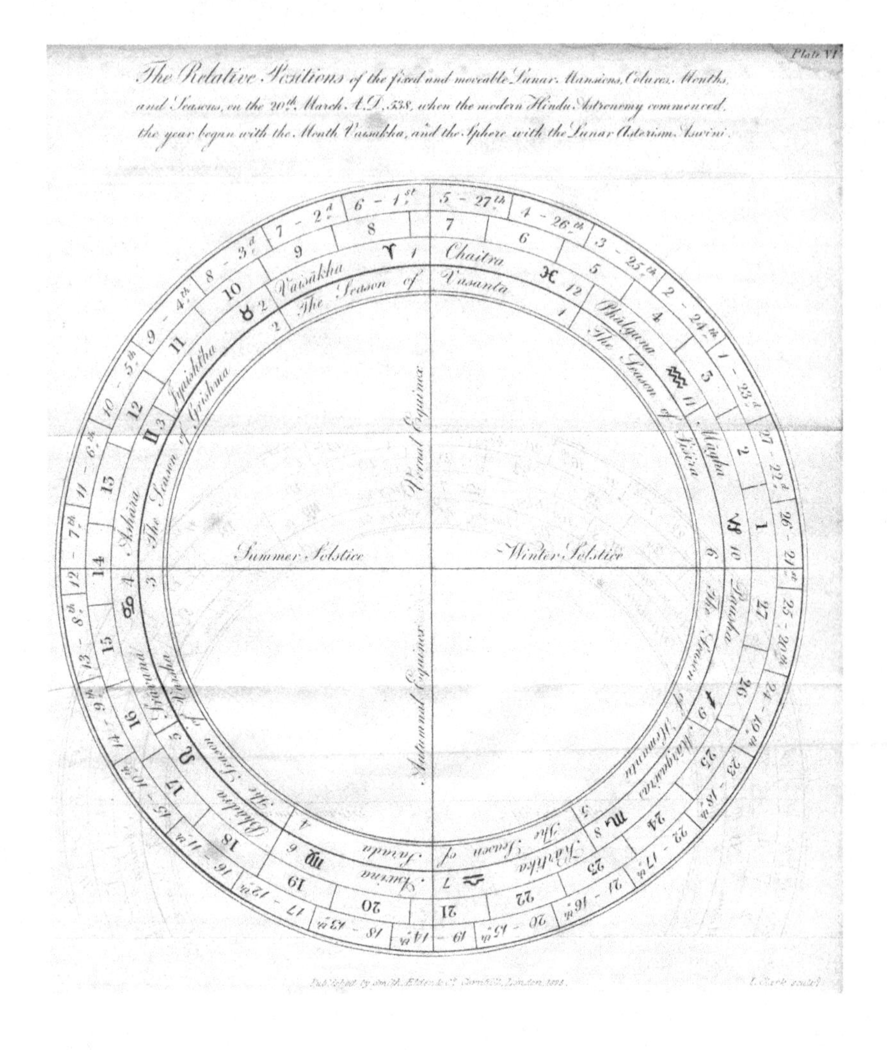

The material originally positioned here is too large for reproduction in this reissue. A PDF can be downloaded from the web address given on page iv of this book, by clicking on 'Resources Available'.

The material originally positioned here is too large for reproduction in this reissue. A PDF can be downloaded from the web address given on page iv of this book, by clicking on 'Resources Available'.

The material originally positioned here is too large for reproduction in this reissue. A PDF can be downloaded from the web address given on page iv of this book, by clicking on 'Resources Available'.

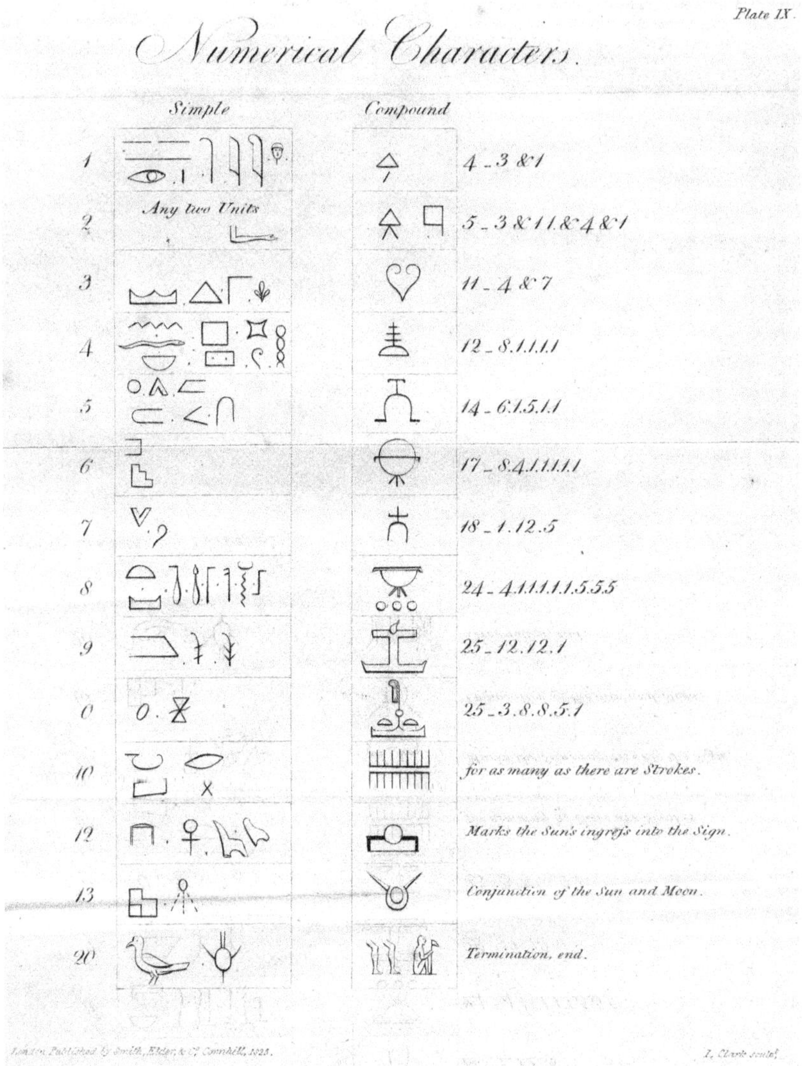

Plate IX.

Numerical Characters.

	Simple	Compound
1		4 _ 3 &1
2	Any two Units	5 _ 3 & 11. & 4 & 1
3		11 _ 4 & 7
4		12 _ 8.1.1.1.1
5		14 _ 6.1.5.1.1
6		17 _ 8.4.1.1.1.1.1
7		18 _ 1.12.5
8		24 _ 4.1.1.1.1.1.5.5.5
9		25 _ 12.12.1
0	0.	25 _ 3.8.8.5.1
10		for as many as there are Strokes.
12		Marks the Sun's ingreſs into the Sign.
13		Conjunction of the Sun and Moon.
20		Termination, end.

London. Published by Smith, Elder, & Co. Cornhill, 1823.

J. Clark sculp.

Printed in the United States
By Bookmasters